71st Annual Meeting of the Labor and Employment Relations Association and LERA@ASSA Meeting 2019

Labor and Employment Relations Association Series

Atlanta, Georgia, USA
4-6 January 2019

Cleveland, Ohio, USA
13-16 June 2019

AA005598

Editors:

Ariel Avgar

ISBN: 978-1-7138-0276-1

Printed from e-media with permission by:

Curran Associates, Inc.
57 Morehouse Lane
Red Hook, NY 12571

Some format issues inherent in the e-media version may also appear in this print version.

Printed with permission by Curran Associates, Inc. (2020)

For permission requests, please contact Labor and Employment Relations Association (LERA) at the address below.

Labor and Employment Relations Association (LERA)
University of Illinois at Urbana-Champaign
121 Labor and Employment Relations Bldg.,
504 E. Armory Ave.
Champaign IL 61820
USA

Phone: (217) 333-0072
Fax: (217) 265-5130

LERAoffice@illinois.edu

Additional copies of this publication are available from:

Curran Associates, Inc.
57 Morehouse Lane
Red Hook, NY 12571 USA
Phone: 845-758-0400
Fax: 845-758-2633
Email: curran@proceedings.com
Web: www.proceedings.com

LABOR AND EMPLOYMENT RELATIONS

ASSOCIATION SERIES

Proceedings of the LERA 2019 Meetings

LERA@ASSA Meeting
January 4–6, 2019, Atlanta, GA
(in conjunction with ASSA/AEA)

and

LERA 71st Annual Meeting,
June 13–16, 2019, Cleveland OH

Ariel Avgar, Editor-in-Chief

LABOR AND EMPLOYMENT RELATIONS ASSOCIATION SERIES
(Formerly the Industrial Relations Research Association)
Proceedings of the Annual Meeting
Annual Research Volume
Perspectives on Work Magazine

Inquiries and other communications regarding membership, meetings, publications, and general affairs of the Association, as well as notice of address changes, should be addressed to the LERA office.

LABOR AND EMPLOYMENT RELATIONS ASSOCIATION
University of Illinois at Urbana-Champaign School of Labor and Employment Relations
121 Labor and Employment Relations Building, MC-504
504 East Armory Avenue, Champaign, IL 61820
Telephone: 217/333-0072 Fax: 217/265-5130 E-mail: LERAoffice@illinois.edu
Website: http://lera.memberclicks.net

CONTENTS

LERA 71st ANNUAL MEETING, JUNE 13–16, 2019, Cleveland, Ohio

VI.
NEW DEVELOPMENTS IN MINIMUM WAGE RESEARCH

Michael Reich, Presiding

VII.

2018 LERA LIFETIME ACHIEVEMENT AWARDS

VIII.
LERA ANNUAL REPORTS

LABOR AND EMPLOYMENT RELATIONS ASSOCIATION SERIES

Proceedings of the LERA 2019 Meetings

LERA@ASSA Meeting
January 4–6, 2019
Atlanta, Georgia

LERA 71st Annual Meeting
June 13–16, 2019
Cleveland, Ohio

OFFICERS OF LERA FOR 2018–2019

PRESIDENT
Kris Rondeau, AFSCME

PRESIDENT ELECT
Dennis Dabney, Kaiser Permanente

PAST PRESIDENT
Harry C. Katz, Cornell University

SECRETARY-TREASURER
Craig Olson, University of Illinois at Urbana-Champaign (2018–January 2019)
J. Ryan Lamare, University of Illinois at Urbana-Champaign (January 2019–present)

EDITOR-IN-CHIEF
Ariel C. Avgar, Cornell University

CHAPTER ADVISORY CHAIR
William Canak, Middle Tennessee State University

EXECUTIVE DIRECTOR
Emily Smith, University of Illinois at Urbana-Champaign

REGIONAL VICE PRESIDENTS
Jim Pruitt, Kaiser Permanente (West Region)
Michele Hoyman, University of North Carolina at Chapel Hill (East Region)
Robert Chiaravalli, Strategic Labor and HR LLC (Mid Region)

EXECUTIVE BOARD MEMBERS
Sylvia Allegretto, Center on Wage & Employment Dynamics
Daniel Altchek, Miles & Stockbridge, P.C.
John Amman, IATSE Local 600
Ezio Borchini, George Washington University
Heather Boushey, Washington Center for Equitable Growth
Bill Dirksen, Ford Motor Company
Virginia Doellgast, Cornell University
Cyndi Furseth, Portland General Electric (ret.)
Kati Griffith, Cornell University
Erin Johansson, Jobs With Justice
Brad Markell, Industrial Union Council (AFL-CIO)
Sheila Mayberry, Arbitrator/Mediator
Jake Rosenfeld, Washington State University, St. Louis
David Weil, Brandeis University

LEGAL COUNSEL
Steven B. Rynecki, von Briesen, Purtell & Roper, S.C.

PAST PRESIDENTS OF LERA

1948—Edwin E. Witte
1949—Sumner H. Slichter
1950—George W. Taylor
1951—William Leiserson
1952—J. Douglas Brown
1953—Ewan Clague
1954—Clark Kerr
1955—Lloyd G. Reynolds
1956—Richard A. Lester
1957—Dale Yoder
1958—E. Wight Bakke
1959—William Haber
1960—John T. Dunlop
1961—Philip Taft
1962—Charles A. Myers
1963—William F. Whyte
1964—Solomon Barkin
1965—Edwin Young
1966—Arthur M. Ross
1967—Neil Chamberlain
1968—George P. Shultz
1969—Frederick Harbison
1970—Douglass V. Brown
1971—George Hildebrand
1972—Benjamin Aaron
1973—Douglas H. Soutar
1974—Nathaniel Goldfinger
1975—Gerald G. Somers
1976—Irving Bernstein
1977—F. Ray Marshall
1978—Charles C. Killingsworth
1979—Jerome M. Rosow
1980—Jack Barbash
1981—Rudolph A. Oswald
1982—Milton Derber
1983—Jack Stieber
1984—Wayne L. Horvitz
1985—Everett Kassalow
1986—Lloyd Ulman
1987—Michael H. Moskow
1988—Phyllis A. Wallace
1989—Joyce D. Miller
1990—Robert B. McKersie
1991—James L. Stern
1992—Ernest J. Savoie
1993—George Strauss
1994—Lynn R. Williams
1995—Walter Gershenfeld
1996—Hoyt N. Wheeler
1997—Francine D. Blau
1998—F. Donal O'Brien
1999—Thomas A. Kochan
2000—Sheldon Friedman
2001—Magdalena Jacobsen
2002—John F. Burton, Jr.
2003—Paula B. Voos
2004—Marlene Heyser
2005—Stephen R. Sleigh
2006—David B. Lipsky
2007—Eileen B. Hoffman
2008—Anthony Oliver, Jr.
2009—Joel Cutcher-Gershenfeld
2010—Eileen Appelbaum
2011—Gordon Pavy
2012/13—David Lewin
2013/14—Lisa Lynch
2014/15—Martin Mulloy
2015/16—Bonnie Castrey
2016/17—Janice Bellace
2017/18—Harry Katz

"Hello, LERA"

KRIS RONDEAU
New England Organizing Project

I've been coming to these meetings for six years now. I come to hear you all talk about the historical transition we are in and, also, for the food. I come to hear the latest progress and innovation from the Kaiser Permanente LMP and from Ford UAW. These stories inspire me and recharge my batteries.

I didn't really figure out what LERA was until I became president of it.

LERA, in all its forms, chapters, constituencies, and publications, is, among its other roles, a friendship organization that brings together a lot of good people who all look at the same problems from different angles. LERA believes in combining professional and student voices, local practitioners, national scholars, and activists. LERA national meeting is, for many, a homecoming. This organization is run by Emily Smith and Bernadette Tiemann, expertly and with love.

I entered the adult work-world researching sleep in a lab but have now dedicated more than four decades to organizing, negotiating, and working beside employees and employers to change work, which needs changing as much as it ever did. Lately, I've been working with unorganized workers … not to form unions, but to help them develop the courage to talk with their employers directly, without any pre-existing framework—like work councils, except without being covered by the National Labor Relations Act. Worse, maybe: a law against that, 8a2 of the NLRA.

In the unions I work with, we don't have a lawyer, because we stopped using them many decades ago when we switched from grievance/arbitration to alternative conflict resolution. So, I called around to many of you for free legal counsel on this build-it-yourself work council idea.

The last call went to LERA member and former head of the NLRB, Wilma Liebman. She was helpful and knowledgeable, and we had a useful conversation. But Wilma threw me off-balance by saying, earnestly, "But, Kris, workers want power, don't they?" Immediately, my stomach was unsettled, and for a long moment I couldn't answer. I've spent every day of my life listening to workers, and I did not know the answer to this question. We all know what we want the answer to be: "Yes. Workers want power. Voice. An identity to be proud of. Of course! Workers want democracy!" But my gut reaction betrayed me. I didn't know what to say.

In our organizing team at my work, we held a special discussion about this question, "Do workers want power?" After two hours of discussion with the most knowledgeable and attuned people I know, we came up with a tentative answer. The best we could come up with was: "Workers don't want to be treated like children." Of this we were sure.

Of course, that can't be right. Each of us wants agency over her own situation. But, for most people, especially working class and working poor, well, you know, they don't work in workplaces that tap into their knowledge! They work in "You're not paid to think" jobs, even in hospitals, universities, and schools. You don't have to work at Amazon or Walmart to be in this kind of trap.

The resulting cynicism can become embedded, passed down through families: Don't ask; don't dare to dream. The working class, in all its forms, newcomers as well as legacy, have every reason to believe the system is rigged. No safety net, no childcare or universal pre-K, a broken and expensive health care system, the everyday fear of being fired from your job. Being working class in America comes with a nagging sense of inferiority that can hurt so much, it can make you mad. It's enough to drive a kind of tribalism. It can make one susceptible to demagogues.

For front-line workers, being able to develop a personal theory and framework for exerting collective power can be an unaffordable luxury. When you're Rita, who works nights so she can pick up per diem shifts during the day …. or José, and you run out of gas on the way to work and get written up for "an occurrence" …. or you're Juana, whose mom is too ill to watch her kids this week: I don't want to sound glib, but these things can have a real inhibiting effect on your creative imagination.

There's still a commonly accepted idea that once a person punches a clock, that person's time belong to the employer. This concept is more than a little soul-destroying. It can really upend your sense of self. It gets into your identity. Do you know how many secretaries (now called administrative assistants), tell me, "In actuality, I am not a secretary. I am a dancer"? A lot of them! I once asked Ahmed, a solidly pro-union worker, to sign his union card. He found that sticky and difficult to do. He said, "If I sign this card, it will be existential proof that I work here."

Some workers use two names: one for work, their real name in their regular life. A longtime Harvard worker, named Harvey, would stand up at any meeting about making work better and say, pretty loudly, "I reserve my right to hate my work." He was pro-union, too. But worker identity in a command-and control workplace is inherently painful.

I was talking with one of my best friends last week about whether or not workers want democracy, and she just laughed. "They're just trying to get through the day without being fired," she said. This stress isn't just undermining a generation of secretaries who are really dancers, it plagues our health. The stress response evolved perfectly to save us from the approaching saber-toothed tigers. Chronic stress, however, is a mismatch condition. It's erased our ability to even get eight hours of nightly rejuvenating sleep. That fact alone undercuts our efforts at health care improvement.

I experience this. You probably do, too. (And you and I? We have very good jobs.) High-demand, low-control jobs make people physically sick. This is well documented in Michael Marmot's wonderful book, *The Health Gap*, on the social gradients of health.

The command-and-control model of management has, for a long time, been a complete disaster, silencing the voices of millions of workers. The power dynamics in most workplaces make workers afraid to speak up, and simply afraid. Their voices are out of practice. Research repeatedly shows that the carrots and sticks used in this model undermine productivity. Carrots and sticks undermine voice. To be honest, I don't see a lot of carrots around. Command-and-control gave us a form of collective bargaining to mirror it. Labor leaders are held at the margins, without space or time to dream about how to make work better. Some days I've felt addicted to transactionalism as the only way we can achieve something.

Mandatory subjects of bargaining, intended to be a good thing, often excluded us from dialogue about work itself. Outside of unions, where nearly 90 percent of all workers reside, the Wagner Act (Section 8a2) banned "company unions" so effectively that for most unorganized workers over the past 84 years, there has been no dialogue, no voice, no everyday negotiations. It's a silence in the workplace that we have all gotten so used to; we can't remember when it was different. In command-and-control workplaces, the best a union could do was protect workers from harm around the edges. Transactionalism is the only item on the menu.

Do workers in these settings dream of democracy? Not really! They dream, if they have a minute, of sitting down. On a good day, they may dream of sticking their feet in a warm ocean. Fortunately, for us, we don't have to kill this form of workplace ourselves, because Millennials are going to do it for us. Guess what? Millennials talk back! Surely, they will put an end to this nonsense.

Still, the idea of unionism was and remains the most inspiring idea I ever heard of. Some of us here—many of us, in fact—gave over our entire lives to this beautiful idea, and to collective bargaining, too, which has so much promise. We recognized the intimacy, the organic culture, of the individual worksite. The potential of it.

Having a union is essential. But if we can't change the way it feels to go to work every day, it's not enough. I know you may be thinking, "Chin up, Kris! Look on the bright side. Collective bargaining is wheezing, sure. But it's just an asthma attack! Or maybe you're just thinking that collective bargaining just doesn't spark joy for me anymore, and I should put collective bargaining into my 'Marie Kondo' reuse bin for the next generation to repurpose."

What can those of us who have loved labor so much do to get it the oxygen it needs? Are there basic values and principles that lay under all structures, the ones we have and the ones to be created?

Whether or not a worker can imagine power at work, she definitely knows that she needs her voice. To imagine democracy, or even power, she must have her voice. Cultivating voice is the essential function of a union. The unions I work with do (obviously, as all unions do) use their voice at the table to negotiate about wages, and health insurance, and retirement. And inclement weather! But we still have what we have. What are the first steps of building something new?

And beyond this, because everything is a negotiation, a navigation, the unions I work with have developed curricula and crafted experiences that teach the techniques and strategies for negotiating in everyday life. And, yes, sometimes skill trumps power.

Organizers are teachers, and they teach transferable skills. Organizers prepare workers for an uncertain path, with twists and turns; we don't steam-roll them a smooth path or a simple message. A simple message is not better than a complex one. With the help of her organizer, a nurse's aide may learn how to negotiate a different start time in order to give care to her ailing mother before she begins her shift. And when she's learned this, she's better prepared to negotiate buying a secondhand car. She's more able to identify intersecting interests when developing an individual education plan with the public school for her special-needs child.

Developing voice is more complex than skills training, more subtle. Voice is a living, breathing thing, and there's no formula. The core activity of unions is one-to-one organizing, the cultivation of voice through individual relationships. Workers change and grow in connection. We all do.

Union organizers—and our members, by extension and through relationships—learn that the purpose of a first conversation is to have a second conversation. And so on. We learn to develop our voices by listening and talking with others. We converse without ceasing. Listening changes the listener. We learn that listening yields real democracy, that it brings something else, even more meaningful: our own beloved community. We become more capable of understanding what it's like to be in our co-worker's shoes.

Without persistent listening, without listening with the third ear, we cannot develop new leaders. As Mary Parker Follett taught us, leading is power-with, not power-over. The natural leaders who pop up first in organizing are often those with the loudest voices. We must take the time to hear the softest voice, the shyest person, the squeak in the floorboards, if we want the communities we build to have wide appeal and staying power.

It is possible to create organizations that are inclusive and safe enough that quality of voice can exist. In this kind of environment, isolated people become connected, passive people become active, and scared people may become brave. When connected and supported, working-class people become ... well, sometimes they do become less cynical. Sometimes they don't. They may still be cynical. But they do carry themselves differently: they're taller. They often learn, without ever saying it explicitly, that they deserve to have their voice listened to. When workers have real conversations in these safe havens, they engage in less either/or thinking; they become more curious, less binary, more subtle, less prone to fight-or-flight, more comfortable in the in-between spaces.

We haven't measured brain size in our unions' members, but I wouldn't be surprised to find that, through this process, the primitive reptilian brain shrinks; the more modern prefrontal cortex gets bigger in relation to the old amygdala. The stress of living with a saber-toothed tiger in the nearby corner office diminishes.

In the early days of our union at UMass Memorial Hospital, our first contract negotiation took longer than many of the women at the hospital's central billing office were comfortable with. A petition began to circulate through their cubicles. "Settle or Strike!" There were posters! Settle or Strike! A frightening thought! We resolved it, through listening and conversation. We got through it. That would never happen now. These workers know how to solve problems and don't let their stress response get the best of them.

As an organizer, even as a union organizer, and even as important as material factors may be, our agenda isn't to change the standard of living for working people. Everyone deserves physical and psychological safety, of course: to have access to quality health care and education, and to be able to afford healthy food--and bread, and roses, too--and to only have to work one job to be able to afford those things.

Living and working standards improve in a sustainable way through deep organizing, through which individuals change their approach to solving problems.

Larger social transformation demands strong growth at the roots. Organizers work to cultivate a capacity for citizenship by fostering it in the local workplace. Transformation, then, comes from a shared sense of responsibility, the idea that making a better society is a project that belongs to all of us, not just the boss, not just an elected official.

It would be easy to think that I've been talking about saber-toothed tigers here as a cute analogy. It could sound kind of like that iconic organizing bumper sticker. You've probably seen this sticker on the back of an old Volvo. It's that one in which the school of little fish is chased by the big-toothed predatory fish. On the sticker, the school of little fish turns around, forms themselves into an even bigger fish, and threatens to eat the predator. "ORGANIZE!" the sticker screams out.

Of course, no real fish actually do that. Not even piranhas … I checked.

Interestingly, if we deconstruct this bad analogy, we uncover a useful scientific truth. The only things that evolve are populations of reproducing individuals. Evolutionary biologist Martin Novak uses mathematical modeling to demonstrate how these evolutionary mechanisms function. His conclusion? That *cooperation* is the master architect of biology.

It would be easy to think that what we do as labor organizers—when we're coordinating recipe swaps, or building community gardens, or telling jokes in our knitting circles, now known all over as "Stitch and Bitch"—it would be easy to think this activity is a quaint aesthetic choice, coincidental to our more "important" work. But if that's what we think, then we have missed a very important thing.

What Novak points out is that culture works just like the rest of life on Earth: cooperation has made possible every advance in cultural evolution, just as it has in biology. When we talk in our unions about how we stir our risotto or whether we grow our tomatoes in the same pots with our basil, when we listen and develop our human voice, we're developing cultural fitness.

If we're going to learn what's next, we must continually learn to cooperate.

This isn't to say that cooperation always works. Competition is obviously an important evolutionary force. But, cancer is what happens when there's a breakdown of cooperation among cells, and it can be fatal.

So, although there's scientific validity to the idea that the world favors those who always cooperate, Novak points out another curious feature to this kind of coordination. The research shows that cooperation never reaches equilibrium. It always falls apart and must be rebuilt. Cooperation can protect us from the fish that's going to eat us, and probably from the saber-toothed tiger, too.

So collective bargaining may be on a ventilator. Labor is out of equilibrium, needs rebuilding. What's next? My old friend (Harvard economist, Secretary of Labor, and past LERA president) John Dunlop used to say, "I don't like a problem you can solve by getting out of bed in the morning. That can't be a very interesting problem. I am only interested in problems that take a decade to solve."

Organizers are teachers, and we're students also. We know that knowledge is perspectival. To understand our own perspective, we must be conversing. To imagine democracy requires us to have the know-how to live in one, at work and in the world. Our voices, our skills of cooperation, our power.

Happily, unions aren't the only organizations with workers' interests at heart. We need to be in constant dialogue with the—yes—Fight for Fifteen … and also with workers' centers, with prison-reform organizations, with artists and humorists and biologists, with the Black Lives Matter, Immigrant and Interfaith and faith-based groups, with broader community-labor coalitions, and with any people we don't know. Unions—and LERA, too—must open all the doors and windows to the new doers. And it's time to step across the street and see what they're talking about over there, too.

Figuring out the path ahead is at least a 10-year problem, which means, by Dunlop's definition, it's interesting. If we are listening as we have been taught, the path ahead will become clear. We will not lose heart. We will organize with love. We can find ways to incorporate what we've done in OUR lives into NEW structures that reflect the ancient values, and we can shepherd our evolution toward a beloved community—a beloved community that, once in a while, we all get to take a vacation from and then dip our toes in some warm ocean water.

Structuring HR and Line Managers' Attitudes Through Negotiation Training: Sensemaking, Sensebreaking, and Sensegiving

BONIFACE MICHAEL
RYAN FULLER
California State University

Abstract

Integrative or principled negotiation binds technical efficiency with moral amicableness and wisdom, typically perfected through practice for lawyers, arbitrators, and mediators. HR and line managers similarly need to develop negotiation-specific attitudinal structures (different from general management) that can balance demands from strategic goals (technical) with common shared concerns (moral) during negotiations. Moral accompaniments to technical insights require faith in first-order facts and higher-order critical insights. Pre-negotiation training can serve as a pre-condition to purposively restructure attitudes toward balancing technical and moral dual demands before arriving at the negotiating table. Such conditioning for negotiations faces collective, cognitive, and evaluative barriers. In this paper, we propose a three-part training method to overcome these barriers. This method structures attitudes among trainees to question their initial assimilation of negotiation principles into their general management orientation and then through dialogical and debriefing processes accommodate critical insights for balancing dual demands in negotiations.

Introduction

Attitudinal structures may be purposively changed to form a pre-condition for negotiations (Walton and McKersie 1965). Such attitudinal structuring has been usefully initiated through negotiation training (Greenhalgh and Lewicki 2015), including pre-bargaining training as at Kaiser Permanente (McKersie et al. 2008) and Saturn (Rubinstein and Kochan 2001) and through executive training such as through the Harvard Business Program on Negotiation (Patton 2009). The co-emphasis on distributive and integrative (Walton and McKersie 1965) efficiency with amicableness and wisdom (Fisher et al. 2011) or subjective with economic value (Curhan et al. 2005) requires structuring attitudes and pre-negotiation training and is a useful first step to initiate attitudinal changes to balance these dual demands during negotiation, different from HR or general management principles.

This paper identifies barriers to critical insights for conflict resolution and offers a process for trainers to break through such barriers. We will next introduce moral and technical demands during negotiation as a duality, then present barriers to balancing this duality during pre-negotiation training, and finally propose a process for breaking through these barriers.

Training for Structuring Attitudes That Balance Technical and Moral Duality

Negotiators' attitudinal structures that balance joint exploration of solutions to common issues without losing sight of principals' interests (Walton and McKersie 1965, pp. 206) is a key objective in pre-negotiation training (Friedman 1994; McKersie et al. 2008). In the quest of this objective, trainers have to educate negotiators about balancing a technical and moral duality embodied in negotiation processes, criteria, and outcomes.

Technical and moral duality. Any negotiation constitutes a process that is a "mixture of distributive and integrative aspects" (Walton and McKersie 1965, p. 206). This process mix may be evaluated through a three-part criterion of "producing a wise agreement, efficiently and amicably" (Fischer, Uri, and Patton 2011, p. 4). Also, there are "social, perceptual and emotional consequences of a negotiation" (Curhan, Elfenbein, and Xu 2005, p. 494) as economic value is created. Successful negotiation therefore implicates the purposive pre-conditioning through training of negotiators' attitudes (Walton and McKersie 1965, p. 266) toward balancing such a moral and technical duality. Technical and economic apportioning of resources, along with finding hidden subjective value for the other side or generated with additional stakeholders, introduces a moral dimension that is difficult to rein together and yet necessary for any successful negotiation training.

Pre-existing attitudinal structures. While the purposeful technical orientation during negotiations comes easily to HR and line managers, balancing the moral consequences does not. The opposite may similarly hold true for union representatives and employees. Institutional practices, collective beliefs, and perceptions (Walton and McKersie 1965) inform attitudes that negotiators bring to the table. Pre-existing psychological and cultural dispositions during negotiations (Weissbein 2000; ElShawnasy 2010) at the table and off stage with constituents (Friedman 1994) pre-dispose negotiators (Curhan and Pentland 2007) toward serving one over the other horn of the duality. Even neurologically, neural networks have been found to suppress moral concerns and openness to ideas due to the overuse of networks that support problem solving and decision making (Boyatzis, Thiel, Rochford, and Black 2017).

Such perceptions and beliefs can usually be overcome with time and practice (Fisher et al. 2011). For HR and line managers (unlike lawyers, arbitrators, and mediators), whose core general management functions do not allow for drawn-out negotiation practice, pre-negotiation training workshops (Walton and McKersie 1965) may purposively create a pre-condition to attitudinal change.

Pre-negotiation training. Most management training and development is treated as goal-oriented expectations for staying relevant and sustainable in a changing business environment that necessitates the pursuit of efficiency and flexibility (Prahlad and Hamel 1990; Cameron and Quinn 2005; Cutcher-Gershenfeld et al. 2006). Even in negotiation training, demands from trainees' functional discipline take precedence (Geenhalgh and Lewicki 2015) and, as a result, further limit their understanding of balancing the technical and moral duality.

According to van Maanen (1983, p. 436), anticipatory learning "covers such matters as expectations, values, skill development, and normative (moral) judgments about the kinds of abilities and performances a person thinks likely to be applicable and rewarded in an imagined new setting." In critical management education, the expectation is that such cognitive and collective learning leads to "raising questions about management and education that are moral as well as technical in nature" (Perriton and Reynolds 2004, p. 65). Pre-negotiation training that emphasizes counter-intuitive integrative and interest-based approaches concomitantly with distributive positional approaches therefore requires critical insightful shifts among trainees' one-sided general management attitudes.

Next in this paper, we identify barriers to identifying technical and moral insights during pre-negotiation training and strategies for overcoming those barriers.

Collective, Cognitive, and Evaluative Barriers

Collective barriers. Pre-negotiation training is underpinned by collective processes such as informational sharing, observation (Nadler, Thomson, and Van Bonne 2003), and joint learning (Patton 2009). Negotiators' formal roles and constituents' informal expectations are powerful contextual conditions that mediate interpersonal interactions during any negotiation (Walton and McKersie 1965), as well as during pre-negotiation training. While appreciation of underlying interpersonal processes may foster learning (Boyatzis, Thiel, Rochford, and Black 2017), formal roles and informal expectations within organizations limit trainees' learning to their functional sub-discipline (Greenhalg and Lewicki 2015). Together these yield a follow-the-leader effect and compliance with one-sided viewpoints that limit the balancing of technical and moral duality.

Cognitive barriers. The challenge in pre-negotiation training is that trainees' minds are not clean slates (Patton 2009) or blank slates (Loes and Warren 2016). Trainees carry an unconscious repertoire of attitudes that influence their perceptions and therefore learning. Presence of existing attitudinal structures (Walton and McKersie 1965) during negotiations predisposes parties and constituents to a functionally or ideologically dominant process—for example, between distributive or integrative processes. Such attitudinal structures are ever present during pre-negotiation training too and limits trainees accommodating the technical and moral duality.

The cognitive underpinning that erodes faith in the facts and stymies understanding and results in incomplete critical reflection is nicely explained by Bean: "No matter what the author really means, students translate those meanings into ideas that they are comfortable with" (2001, p. 135). Pre-negotiation training suffers from cognitive barriers that hastily propel trainees to early conclusions and concurrently paralyze them from moving to new, second-order interpretations necessary to accommodate the technical and moral duality.

Evaluative barriers. According to Susskind and Coburn (2000), negotiation training is marked by "last-day-of-the-term" (308) evaluations, informal post-program feedback, journal entries, examinations, or word-of-mouth reactions. Without having trainees and instructors review and hear each others' perspectives, training is incomplete. Making meaningful sense of the technical and moral duality is an important concluding step for training evaluation. According to Patton, "Students trained through a lifetime of finding the 'right answers' would quickly hone in on the 'theme of the day' ('Oh, it's Options Day') and produce relevant insights. But those insights would seem to have no lasting impact on their awareness or behavior anytime later" (2009, p. 487).

The traditional approach to learning in executive training or within the classroom involves some terminal reflection, discussion, and presentations followed by end-of-training evaluation. In the classroom, student feedback and assessment through examination and term papers is the primary mode of learning assessment (Bean 2001). However, large class sizes coupled with terminal assessments and papers limit the extent to which students' paralysis is overcome and original, critical insights are revealed and enriched throughout the process. Oftentimes, trainees will regurgitate the instructor's critical insights during debriefing.

In the next section, we present a case study that illustrates collective, cognitive, and evaluative barriers to accommodating technical and moral duality encountered during pre-negotiation training and strategies for overcoming them.

Case Study: Two Tales of Negotiation Training

The case study presented here is based on three upper-division undergraduate conflict management and negotiation sections. Of the students, 37.5% had between 1 to 5 years of work experience, 35% had between 10 and 15 years, and 7.5% students had 15 years.

Course Mechanics: Duality, Facts, and Interpretations

Scholarly articles on business strategy, competing strategic values, work organizations, and three-dimensional dealcrafting, along with tracking a Stanford business negotiation video on leasing a baseball stadium and episodes of the television series "The Good Wife" framed the technical side of negotiation. Principled and interest-based negotiation approaches framed the moral side of negotiation.

A scaffolding approach iteratively moved learning from faith in the facts to higher-order interpretations. Faith in the facts was developed through 11 before-class weekly facilitations and accompanying responses to scholarly readings on a discussion board through self-selected roles. Weekly course reading materials were released on a Thursday, with facilitators' posts due by Saturday and responses due by Monday. This organic creation of facilitator and respondent roles set the stage for in-class discussion of weekly topics and receiving weekly grades. To facilitate technical and moral balancing, different components ended with debriefing and synthesizing material from written assignments for in-class discussions.

A Tale of Barriers

Early in the course (weeks 1 through 3), even though the course mechanics were enacted and students engaged in dialogue and received weekly grades, they still experienced collective and cognitive barriers. These barriers kept students' critical insights mostly one sided, either technical or moral, with minimal recognition of duality.

Collective barriers. Respondents followed the lead of facilitators' posts even if the facilitators' interpretation of facts was incorrect. For example, on Prahlad and Hamel's seminal work on core competencies (1990), a facilitator interpreted the statement that "the article illustrates the competition between Western and Japanese auto industries." This resulted in the article being interpreted as a clash of management cultures, similar to the follow-the-leader approach in the workplace. Early student responses in that week appeared to be cued by the facilitator's Japanese versus Western interpretation: "Like the situation between Japanese and Western companies; Western companies were behind Japanese companies (and continued to fall behind)."

Pre-class reading habits in the classroom mirrored those of trainees in the real world. Students quickly read, summarized, and critically analyzed articles as the due date for submission loomed, quite similar to HR and line managers preparing for training in between their many other daily tasks. This often leads to factual inaccuracies or underdeveloped interpretations of the original author's scholarly work.

Cognitive barriers. Differences in use of meanings of words are essentially a reflection of pre-existing cognitive paradigms within individuals (Patton 2000), and learners misinterpret facts when they equate them with terms that they find comfortable (Bean 2001). Cameron and Quinn's (2005) competing values framework reviewed early in the course provided such an example of this forcing of familiar meaning onto new concepts and assimilating new concepts into existing paradigms. Facilitators and respondents consistently equated competing values about strategic effectiveness criteria to the more familiar notion of shared values in four different cultural types that are also present in Cameron and Quinn's model.

Although trainees sensed meaning, they fell short of fully comprehending conceptual differences. For example, while examining arbitration as an alternative dispute resolution mechanism, students collectively forced their newly gained understanding of the appealing approach of principled negotiation onto the more formal and binding decisiveness of neutral, third-party arbitration and revealed inchoate insights.

An Alternate Tale of Breaking Through the Barriers: Sensemaking, Sensebreaking, and Sensegiving

The method adopted in the classroom yielded an alternate three-part process involving sensemaking (Weick 1996 a) through reforming the context (Greenwood and Hinings 1996); sensebreaking (Pratt 2000) that creates meaning voids and questions existing perceptions; and sensegiving (Gioia and Chittipeddi 2001) through interactive dialogical pathways (Grant and Marshak 2011; Heckscher and Adler; 2006) that leads to accommodation.

Sensemaking. According to Weick (1996a), sensemaking is an ongoing process of dominant confrontations with the environment and collective enactments. It involves identity construction influenced by contextual frames of reference that directs interpretation through the creation of meaning. Identity's doggedness within a given environmental reference frame and dominant behavioral paradigms similarly raises collective barriers in the training classroom, to reformation commitment (Greenwood and Hinings 1996; Putnam 2010) necessary for critical insights in conflict management and negotiation training.

Sensemaking can be usefully initiated by placing trainees in an environment in which they have to share with others the takeaways that they have assimilated into their own interpretations. The training delivery mechanics recast the earlier flocking patterns of students writing and sharing their first-order facts publicly on discussion boards into taking a stand on higher-order interpretations (partially developed or inaccurate).

Recasting the context, as a first step, is a time consuming and coordination-laden act that requires instructors not just to collate and make training material accessible before the program but also to reform trainees' commitments to a new value to take a stand before they meet in class, be it inaccurate facts or initial and partial assimilated insights. This important first step needs to be continued by strengthening faith in the facts and building valid critical insights by breaking past cognitive barriers, which is presented next in this case study.

Sensebreaking. Patton (2000) describes the use of symbolic enactments by actually tearing up a physical copy of *Getting to Yes!* in the classroom as a way to break past trainees' existing cognitive paradigms and folklore. Such sensebreaking has been found to create meaning voids among trainees, as in the case of Amway (Pratt 2000), which is a necessary step for motivating them to seek positive new meaning that more validly fits organizational needs. Trainees' first sensemaking, often assimilated into their existing cognitive maps, is questioned when these individual early conclusions conflict with alternate legitimate insights and shake faith in the facts and the interpretations they yield.

Early in the term, at its most basic, students' facts describe the concepts; they do not define and explain the concepts and their links within and across weekly readings. The instructor's weekly grading, in-class dialogue, and debriefing helped break past these habits and build faith in the facts and improve on valid interpretations.

Even the tendency to follow the leader began to unravel and was deflected with alternate valid interpretations. In one case, later in the course a facilitator raised a question, "Why is implantation more important than a simple yes?" Instead of being led by the question, respondents broke away from implantation and focused on implementing a deal, which was the topic for the week. One replied, "Implementation matters anytime there is something that we must do to get the benefit of our bargain, and to extract that value over time." Another responded simply, "Implementation is important as it serves as a way for parties to follow through in the agreement." Another concluded, "Implementation is the process that turns strategies and plans into actions in order to accomplish objectives and goals."

This breaking away and proliferation of diverse insights also needs one final enactment of processes that lead to critical insights to accommodate the technical and moral duality in negotiations. This final step involves the students' and instructor's joint debriefing that begins to accommodate new meaning to fill the voids created; that process is presented next in this case study.

Sensegiving. The creation of meaning voids results in a proliferation of multiple frameworks that may be usefully shaped by offering a preferred alternate template (Lewin 1951). This sensegiving (Gioia and Chitipedia 2001) process moves the initial sensemaking that has been assimilated within trainees' existing knowledge and the meaning void experienced through sensebreaking toward a plausible interpretation that unravels a line of sight to the technical moral duality of negotiation.

In the classroom, a key debriefing exercise involved students who facilitated and responded to a specific topic presenting their discussion board themes and dialoguing with other students, which was also facilitated by the instructor. The written discussion threads provided a potent reference point in this debriefing activity. In-class arbitration and negotiation simulations were followed up with debriefing sessions in which students shared their own insights, which were again facilitated by the instructor. Students also submitted an individual compilation of tools for conflict management or negotiation, which was reviewed and commented by the instructor for students to revise and finalize. These debriefing engagements, including the preliminary written discussion threads, were informed by a publicly available learning rubric and instructor's prompt for conceptual accuracy, cross-conceptual linkages, practical application, and writing proficiency.

Also, in-class negotiation simulations, stretched out over 13 weeks, let students assume roles of project owners and vice presidents rather than consultants (simulating employer–employee relations), and accommodative sensemaking became visible in action. For example, while negotiating on terms and conditions for working together on a group project, an impasse was reached on an assessment scale. While one group wanted a 10-point scale, another group wanted a 5-point scale. Negotiators finally agreed on a 5-point scale that was rated in increments of 2.

This tripartite alternate tale for conflict management and negotiation training involves students and instructors in equally important roles. The instructor plays a key role by setting the stage, providing weekly feedback on students' contributions, and sharing his/her own point of view without letting it become the default correct template that students assimilate and regurgitate. Interactive dialogue among students, facilitated by the instructor, helps students develop and sharpen their interpretation of facts. Verbal and written debriefing throughout the course helps draw a line of sight between the technical and moral dimensions underpinning conflict management and negotiation training.

Implications and Concluding Comments

Attitudinal structures that balance conflicting goals with common shared concerns (Walton and Mckersie 1965); that recognize the generation of subjective as well as economic value (Curhan et al. 2005); and that oriented toward efficient processes, as well as amicable and wise outcomes (Fischer et al. 2011) are key learning objectives sought by negotiation trainers and trainees. This paper helps trainers and trainees cope with comprehending these dual demands as a technical and moral duality faced by HR and line managers during negotiations.

Unlike lawyers, arbitrators, and mediators who perfect their skills through practice, HR and line managers have to often negotiate on the fly alongside their other general management responsibilities. Faced with limited practice time, pre-negotiation training is a first step to purposively develop conditions for balancing technical and moral duality necessary for negotiating. This paper identifies collective, cognitive, and evaluative barriers to the introduction of attitudinal structures for balancing technical and moral duality during negotiations. Together these barriers erode factual accuracy and limit the higher-order critical insights necessary for balancing the moral and technical demands of negotiations.

To overcome these barriers, this paper offers an alternate three-part pre-negotiation training delivery method that places trainers and trainees in a joint exercise of sensemaking, sensebreaking, and sensegiving. This method sharpens factual accuracy and brings together technical and moral duality in trainees' critical insights of negotiation. Recasting the environmental context through technology, such as discussion boards, within and beyond the walls of the classroom, facilitates exchanges between the dominant first and often inaccurate viewpoints and insightful alternate viewpoints. While instructor-led scenarios, in-class simulations, and interactive dialogue among trainees break habits and one-sided viewpoints, multi-modal purposeful written and verbal debriefing jointly by trainers and trainees strengthens and transforms the learning process.

This three-part method also has implications for negotiation trainers and trainees. Possibly some of the most salient models for attitudinal structuring (Walton and McKersie 1965) involve role reversals (McKersie et al. 2008; Friedman 1994; Heckscher and Hall 1994) and experiential simulations (Susskind and Corburn 2000). The training delivery method proposed here may be usefully replicated in short-term executive or continuing education training programs as well.

This method omits consideration of students' learning styles, such as simulation, analogy, and didactic information sharing, all of which were present in some form or shape in the courses examined. However, this method introduces the salience of multiple modes of writing as a debriefing medium for imparting conflict management and negotiation training. Continuing development of this method should examine and link different training modes with trainees' learning styles.

References

Bean, J.C. 2001. *Engaging Ideas: The Professors Guide to Integrating Writing, Critical Thinking, and Active Learning in the Classroom.* San Francisco, CA: San Francisco: Jossey-Bass.

Boyatzis, R.E., K. Thiel, K. Rochford, and A. Black. 2017. "Emotional and Social Intelligence Competencies of Incident Team Commanders Fighting Wildfires." *Journal of Applied Behavioral Science* 53 (4): 498–516.

Cameron, K.S., and R.E. Quinn, E. 2005. *Diagnosing and Changing Organizational Culture Based on the Competing Values Framework.* Hoboken, NJ: John Wiley and Sons.

Coleman, P.T., and Y.Y. Joanne Lim. 2001. "A Systematic Approach to Evaluating the Effects of Collaborative Negotiation Training on Individuals and Groups." *Negotiation Journal* 17 (4): 363–392.

Curhan, J., H. Elfenbein, and H. Xu. 2005. "What Do People Value When They Negotiate? Mapping the Domain of Subjective Value in Negotiation." IDEAS/RePEc Working Paper Series.

Curhan, J.R., and A. Pentland. 2007. "Thin Slices of Negotiation: Predicting Outcomes from Conversational Dynamics Within the First 5 Minutes." *Journal of Applied Psychology* 92 (3): 802.

Cutcher-Gershenfeld, J., S. Sleigh, and F.K. Pil. 2006. "Collective Bargaining: Keeping Score on a Great Institution." In *Contemporary Issues in Employment Relations.* Champaign, IL: LERA.

Elshenawy, E. 2010. "Does Negotiation Training Improve Negotiators' Performance?" *Journal of European Industrial Training* 34 (3): 192–210.

Fisher, R., W. Ury, and B. Patton. 2011. *Getting to Yes: Negotiating Agreement Without Giving In*, 3rd ed. New York: NY, Penguin Books.

Friedman, R.A. 1994. *Front Stage, Back Stage: The Dramatic Structure of Labor Negotiations.* Cambridge, MA: MIT Press.

Gioia, D.A., and K. Chittipeddi. 1991. "Sensemaking and Sensegiving in Strategic Change Initiation." *Strategic Management Journal* 12 (6): 433–448.

Grant, D., and R.J. Marshak. 2011. "Toward a Discourse-Centered Understanding of Organizational Change." *Journal of Applied Behavioral Science* 47 (2): 204-–35.

Greenhalgh, L., and R.J. Lewicki. 2015. "Evolution of Teaching Negotiation: The Legacy of Walton and McKersie." *Negotiation Journal* 31 (4): 465–476.

Greenwood, R., and C. Hinings. 1996. "Understanding Radical Organizational Change: Bringing Together the Old and New Institutionalism." *Academy of Management Review* 21 (4): 1022.

Heckscher, C., and L. Hall. 1994. "Mutual Gains and Beyond: Two Levels of Intervention." *Negotiation Journal* 10 (3): 235–248.

Heckscher, C.C., and P.S. Adler. 2006. *The Firm as a Collaborative Community. The Reconstruction of Trust in the Knowledge Economy.* New York, NY: Oxford University Press.

Kolb, D.A. 1984. *Experiential Learning: Experience as the Source of Learning and Development.* Englewood Cliffs, NJ: Prentice-Hall.

Lax, D.A., and J.K. Sebenius. 2002. "Dealcrafting: The Substance of Three-Dimensional Negotiations." *Negotiation Journal* 18 (1): 5-28.

Lewin, K. 1951. *Field Theory in Social Science.* New York, NY: Harper and Row.

Loes, C., and D. Warren. 2016. "Using Observational Learning to Teach Negotiation." *Journal of the Academy of Business Education* 17: 296–310.

Loewenstein, J., and L. Thompson. 2000. The challenge of learning. Negotiation Journal 16 (4):399 - 408.

McKersie, R.B., T. Sharpe, T.A. Kochan, A.E. Eaton, G. Strauss, and M. Morgenstern. 2008. "Bargaining Theory Meets Interest-Based Negotiations: A Case Study 1." *Industrial Relations: A Journal of Economy and Society* 47 (1): 66–96.

Nadler, J., L. Thompson, and L. Van Boven. 2003. "Learning Negotiation Skills: Four Models of Knowledge Creation and Transfer." *Management Science* 49 (4): 529–540.

Patton, B. 2009. "The Deceptive Simplicity of Teaching Negotiation: Reflections on Thirty Years of the Negotiation Workshop." *Negotiation Journal* 25 (4): 481–498.

Perriton, L., and M. Reynolds. 2004. "Critical Management Education: From Pedagogy of Possibility to Pedagogy of Refusal?" *Management Learning* 35 (1): 61–77.

Prahalad, C.K. 1993. "The Role of Core Competencies in the Corporation." *Research–Technology Management* 36 (6): 40–47.

Prasad, P., and P.J. Caproni. 1997. "Critical Theory in the Management Classroom: Engaging Power, Ideology, and Praxis." *Journal of Management Education* 21 (3): 284–291.

Pratt, M.G. 2000. "The Good, the Bad, and the Ambivalent: Managing Identification among Amway Distributors." *Administrative Science Quarterly* 45 (3): 456-493.

Putnam, L.L. 2010. "Negotiation and Discourse Analysis." *Negotiation Journal* 26 (2): 145–154.

Rubinstein, S.A., and T.A. Kochan. 2001. *Learning from Saturn: Possibilities for Corporate Governance and Employee Relations.* Ithaca, NY: ILR Press.

Susskind, L. 1996. *Dealing with an Angry Public: The Mutual Gains Approach to Resolving Disputes.* New York, NY: Free Press.

Susskind, L.E., and J. Corburn. 2000. "Using Simulations to Teach Negotiation: Pedagogical Theory and Practice." In *Teaching Negotiation: Ideas and Innovations*. Cambridge, MA: PON Books.

Van Maanen, J. 1983. "Golden Passports: Managerial Socialization and Graduate Education." *Review of Higher Education* 6 (4): 435–455.

Walton, R.E., and R.B. McKersie. 1965. *A Behavioral Theory of Labor Negotiations: An Analysis of a Social Interaction System*. New York, NY: McGraw-Hill.

Weick, K. 1996a. "Prepare Your Organization to Fight Fires (*Young Men and Fire* by author Norman Maclean Helps Teach Lesson in Corporate Structure)." *Harvard Business Review* 74 (3): 143.

Weick, K.E. 1996b. *Sensemaking in Organizations*. Thousand Oaks, CA: Sage.

Weissbein, D. 2000. "Improving Training Effectiveness Through Motivation: Creating a Psychological States Intervention." Ph.D Thesis, Michigan State University.

Market Power, Finance Wages and Inequality

WENTING MA
University of Massachusetts, Amherst

Abstract

Increasing industry concentration has raised concerns that declining competition among firms for labor has led to slow wage growth. However, the financial sector has been an exception. I find that finance wages have increased by almost three times the increase in non-finance wages, despite similar trends in market concentration. Using administrative data from the U.S. Census, I construct measures of firm-specific market power and show that higher market power is associated with significantly higher wages in finance than in non-finance. I provide evidence that rent-sharing plays an essential role in driving the more pronounced effect of market power on finance wages for two reasons. First, financial firms with higher market power can extract relatively higher rents to share. Second, financial firms give a relatively higher share of rents to workers, especially high-skill workers, due to relatively higher worker bargaining power. As rents are disproportionately distributed to high-skill workers, financial firms with higher market power are associated with relatively higher within-firm inequality.

Introduction

Recent literature documents an important trend in U.S. product markets: increasing concentration. Grullon et al. (2017) show that 75% of U.S. industries have become more concentrated since the 1990s, and the average firm is almost three times larger. One potential concern with this rise in industry concentration is that it reduces workers' employment options, and thus gives employers the ability to lower wages (Manning, 2011; Stiglitz, 2017; Benmelech et al., 2018).[1] The finance sector, however, has been an exception. I observe that from 1990 to 2008, real wages increased by 23.38% in finance but only 8.85% in non-finance. In the same period, the degrees of industry concentration as measured by the Herfindahl-Hirschman index (HHI) increased by roughly 40% in both finance and non-finance. These novel findings indicate that market concentration may not have the same wage dampening effect in the financial sector, suggesting a more nuanced understanding of the effects of concentration on wages and the finance wage premium.

Why does concentration impact finance wages differently? In this paper, I argue that rent-sharing plays an essential role in driving the difference. Industry concentration increases the market power of firms, especially larger firms within a given industry. An increase in firms' market power not only increases firms' labor market monopsony power to lower wages by decreasing competition for hiring workers, but also increases firms' product market monopoly power by decreasing competition for selling products or buying inputs. With higher product market monopoly power, firms can charge higher markup and thus extract higher rents. Wages rise when firms share these rents with their workers.[2] Therefore, higher market power is associated with two competing effects on wages: rent-sharing effect and labor monopsony effect. In finance, the rent-sharing effect appears to dominate the labor monopsony effect, and the net effect is relatively stronger. To support this argument, I propose and provide empirical evidence on two non-mutually exclusive mechanisms. First, financial firms with higher market power can extract relatively higher rents, and thus they have more to share with employees. Under this scenario, financial firms' profitability responds more positively to the increase in firm market power relative to non-financial firms. Second, financial firms may have to give a larger fraction of rents to employees, especially high-skilled workers, due to higher worker bargaining power. Under this scenario, the wages of high-skilled workers in finance respond more positively to an increase in firm market power as compared to those in non-finance.

Using administrative micro-level data from the U.S. Census, this paper first examines the relation between wages and industry concentration, and whether the relationship differs in finance. I find that concentration measured by HHI is negatively correlated with wages in non-financial industries. By contrast, HHI is positively correlated with wages in finance, and the positive correlation is statistically significant at 1% level. These findings are robust to various measures of industry concentration, indicating that industry concentration has different implications in finance.

While industry concentration means higher market power for firms remaining in the industry on average, previous studies argue that concentration disproportionately benefits larger players within industries. Relative to small players in a given industry, larger firms have higher market power to raise prices or lower input prices and thus make higher rents. In other words, market power should be firm-specific and dependent on the firm's market share, implying that rents increase with firm size within industries (Shepherd, 1972; Porter, 1979; Smets and Wouters, 2007; Boar and Midrigan, 2019). With more rents, firms with higher market power can afford higher wages on average as compared to firms with lower market power within a given industry. Using this within-industry variation, I next conduct firm-level cross-sectional analysis to examine how the relationship of firm-specific market power and wages in finance is different from that in non-finance.

As the baseline, a firm's market power is defined as its employment share in its industry.[3] The main result of this paper is that higher firm market power is associated with significantly higher wages in finance than that in non-finance. This result holds when controlling for a battery of factors that are likely to drive the heterogeneity of wage-setting behaviors across firms. Also, this result holds for defining firms' market by either two-digit or three-digit Standard Industry Classification (SIC). The difference between finance and non-finance is still significant if I instead look at median wages, average wages adjusted by cost of living, and average wages of male or female workers.

I next provide evidence for two non-mutually exclusive mechanisms that can explain why the positive relationship between firm market power and wages is stronger in finance. First, I argue that financial firms with higher market power can extract relatively higher rents as compared to non-financial firms. Market power may be particularly valuable in finance due to a stronger belief in "too big to fail," or less geographical restrictions for allocating resources. For example, the "too big to fail " argument suggests that larger financial firms are more likely to be supported by the government when they face potential failure, and thus they can borrow at a much lower cost than smaller financial firms who must borrow based on their creditworthiness (Baker et al., 2009). Because financial firms can extract relatively higher rents with higher market power, they have more to share with workers. In support of this mechanism, I find that firms with higher market power exhibit higher profitability measured by return on assets (ROA) in finance relative to non-finance.[4]

To further investigate the sources of the additional profitability created by firm market power in finance, I decompose ROA into two components: the Lerner Index and the Asset Utilization ratio. The Lerner Index (or price-cost margin) has been widely used in the literature as a proxy for the extent to which prices exceed marginal costs (e.g., Muller et al., 2017; Aghion, 2005; Grullon et al., 2017). The asset utilization ratio captures operational efficiency (Grullon et al., 2017). Results show that financial firms can generate relatively higher profitability with higher market power because they are able to charge significantly higher price-cost margins relative to non-financial firms.

Second, compared to non-financial firms, financial firms may have to give a higher proportion of their rents to workers due to higher worker bargaining power. This mechanism should apply especially to high-skilled workers in finance because they are more likely to be matched with larger scale tasks, or high external visibility of their performance increases the probability of being poached. I find that a one standard deviation increase in firm market power is associated with 1.7% higher average wages for high-skill workers in non-financial firms, whereas the effect is 6.37% in finance, which can be translated into $1468 per quarter. The results are robust to various definitions of high-skill workers.

As a consequence of rents being disproportionately distributed to high-skilled workers within firms, I find that financial firms with higher market power exhibit higher within-firm inequality, especially among male workers, as compared to non-financial firms.

The final part of my study conducts a comprehensive set of robustness analysis to eliminate alternative explanations. First, I conduct firm-local labor market level analysis to examine whether the difference in finance arises because non-financial firms are more likely to compete locally and measuring market power across national markets underestimates the firm's ability to extract rents. Results at firm-local labor market-level are qualitatively similar to those at firm-level and do not support this alternative explanation. Second, I re-define market power using sales data to make sure market power is not more valuable in finance because market power measured by domestic employment captures characteristics that are more valuable in finance than in non-finance. Third, finance wages still have the highest sensitivity to market power when compared with wages in non-financial industries facing low import exposure. This result helps alleviate the concern that market power measured by domestic share is less valuable in non-finance because non-finance firms face higher import competition. Lastly, I conduct individual-level analysis while controlling for individual time-varying and time-unvarying quality. I continually find higher firm market power is associated with significantly higher wages in finance as compared to non-finance, which alleviates the concern that my results could be explained by sorting effect based on worker quality.

My paper builds on several bodies of literature. First, it builds on the literature on market concentration and its potential effects on wages. Much of this literature focuses on non-finance sectors, and a discussion on *how* and *why* concentration plays a differential role in finance wages has been missing. Weiss (1966) shows that wages are higher in more concentrated industries. However, the relationship is no longer significant and positive after controlling for personal characteristics, and the results only hold within manufacturing, transportation, and utility industries. Landon (1970) finds higher concentration in the newspaper industry is associated with lower wages. More recently, Autor et al. (2017) document industries where concentration increases the most have the sharpest decline in the labor share. Labor share is measured by payroll-to-sale ratio, which is not equivalent to real wages. Benmelech et al. (2018) show that employer concentration increases firms' labor monopsony power, thus lowering wages in manufacturing industries. Due to data availability in labor productivity, they only focus on manufacturing industries. This study contributes to the literature by uncovering the differential role of market power on finance wages. I further provide evidence that rent-sharing can help explain the difference. Meanwhile, this study also contributes to the literature by showing that higher firm market power has different effects on the wages of different skill-level workers, thus contributing to higher within-firm inequality.

This study complements the literature that seeks to understand the substantial wage premium in financial industries (Philippon and Reshef, 2012; Boudtanifar et al., 2018; Bohm, Metzger and Stromberg, 2018; Axelson and Bond, 2015). Consistent with previous literature, I confirm finance wages are on average higher than non-finance wages using data from the U.S. Census. While previous literature shows that worker- and industry-level characteristics contribute to high wages in finance, these factors cannot fully explain the surge of finance wage premium (Philippon and Reshef, 2012; Bohm, Metzger and Stromberg, 2018). Building on previous literature, my paper provides a more nuanced understanding of finance wage premium through the lens of firms. Using U.S. employer-employee matched data, I show that higher market power is associated with higher wages in finance because market power is associated with a stronger rent-sharing effect in finance. My results suggest that firm market power can help explain the finance wage premium.

This paper also builds on the literature that investigates factors affecting within-firm inequality. Muller, Ouimet, and Simintzi (2017) show that, on average, larger firms have higher pay inequality using UK data. Consistent with their finding, my results show firms with relatively larger sizes in a given industry exhibit higher pay inequality. My study further shows that firms that are relatively larger in financial industries are associated with even higher within-firm inequality because rents are disproportionately distributed to finance high-skilled workers who have relatively higher worker wage bargaining power. Ma, Ouimet, and Simintzi (2018) document that mergers and acquisitions (M&A) act as the catalyst for firm technology adoption, which in turn leads to an increase in inequality within target establishments. Instead of looking at how firm reorganization affects the wage distribution within the firm, my paper shows that the variation in firm market power can explain the heterogeneity of within-firm inequality.

The remainder of the paper is organized as follows. Section 2 describes data sources and sample constructions. Section 3 presents empirical analysis on the relationship between market power and wages and how it is different in finance. Section 4 presents empirical evidence on mechanisms. Section 5 presents analysis on the relationship between market power and within-firm inequality. Section 6 conducts additional robustness analysis to eliminate alternative explanations. Section 7 discusses external validity, and Section 8 concludes.

Data

In this section, I start with reviewing multiple data sources used in this study and describing how I combine them to construct the baseline sample. I then describe how firms' industry and financial firms are defined. At the end of this section, I construct the measures of industry concentration and firm market power.

Data Sources

The analysis in this paper combines data from three confidential databases maintained by the U.S. Census Bureau: 1) the Longitudinal Employment-Household Dynamics database (LEHD); 2) the Longitudinal Business Database (LBD); and 3) the Business Register(BR). I also link firm financial statement data from Compustat to firms in LBD through a Census internal bridge.

I use the LEHD to obtain information on firms' wage patterns and workforce composition. The LEHD is an employer-employee matched database which tracks employees and their wages with various employers on a quarterly basis.[5] Individual wages reported in the LEHD include all forms of compensation that are immediately taxable, including bonuses and exercised stock options which take a heavyweight in finance sector pay.[6] The LEHD also reports age, gender, and education level of each employee.[7] As workers in this program can be linked to their employers, it allows me to track wage distribution and workforce composition within each employer. The data start in 1990 for several states and coverage of states increases over time. The data coverage ends in 2008. This project has access to 31 states.[8] I map these states in Appendix Figure 3. While I do not observe data for all states, I observe almost 100% of private employment for any state in the program. I discuss the consequence of omitting states in section 7.

I supplement the information in the LEHD with firm-level information on employment and industry from the LBD.[9] This database tracks all US business establishments on an annual basis. An establishment is any separate physical location operated by a firm with at least one paid employee. The LBD includes information on industry, the number of employees and total payroll at each establishment. Also, the LBD contains a unique firm-level identifier which longitudinally links establishments that are part of the same firm. As the LBD tracks all establishments in the U.S., it allows me to measure total domestic employment for each firm and industry by aggregating employment across establishments. In section 2.3, I will discuss how I utilize establishment employment and industry to define firms' industry and identify financial firms.

To conduct mechanism tests in section 4, I collect firm financial information on earnings before interests, taxes, depreciation and amortization (EBITDA), total sales, and total assets from Compustat.

Lastly, I obtain sales data for firms from the Business Register (BR) database to conduct robustness checks in section 5.[10] This database collects business sales, including total revenue from selling products, interest income and gross rents, from the Internal Revenue Service (IRS). It tracks firms from all 50 states and the District of Columbia on an annual basis. A key advantage of the sales data from BR over standard firm-level databases such as Compustat is that they comprise both private and public listed firms.

Sample Construction

The baseline sample is at firm-year-quarter-level and spans from 1990 to 2008. To obtain firm wage patterns and workforce composition, I start with linking filtered employee-level data from the LEHD to firm identifiers in the LBD through federal employer identifier (EIN), and then I aggregate employee-level data to firm-level in each year-quarter.[11]

Specifically, I restrict my attention to full time workers in the LEHD by only including workers aged between 16 and 65 years old and by excluding employee-quarter that earned less than 80% of the 1990 federal minimum wage following Philippon and Reshef (2012), where wages are converted to constant 2001 dollars.[12] To ensure that I am not observing quarters in which an employee was only partially employed at a given firm, I only keep employee-firm quarters where I observe a full quarter of employment at the firm prior to and post that quarter.[13]

I then link the filtered employee data to firms in the LBD to construct firm-year-quarter-level measures of wage patterns and workforce compositions.[14] Since self-employment may have different wage setting behavior, I exclude firms who only have one paid employee to minimize the possibility of picking up self-employment. I define all the variables used in my analysis, in more detail, in the Appendix.

Define Firm Industry and Financial Firms

Throughout the study, I use the 1987 Standard Industry Classification (SIC) codes to define markets. To define a firm's market (termed "firm industry"), I use establishment-level information on SIC and employment from the LBD.[15] For a firm owning only one establishment, its industry is defined by its establishment's industry. For a firm owning multiple establishments spanned multiple industries, the firm is classified in an industry where it has more than 50% of its employment. I drop a negligible percentage of firms who span in multiple industries, but their employment share in each industry is less than 50%. Following these rules, a firm's industry is defined by a 2-digit or 3-digit SIC code.[16]

Following Philippon and Reshef (2012), financial industries include depository institutions (except central reserve depository institutions), non-depository institutions, security and commodity brokers, insurance carriers, insurance agents, brokers and service and holding and other investment offices. Non-financial industries include all other non-farm private industries. For a corporate to be a financial firm, its industry must be in one of the financial industries. Similarly, a firm is classified as a non-financial firm if its industry is in one of the non-financial industries.[17] Based on this classification rules, financial firms account for approximately 4.3% of my baseline sample, which is similar to the statistics reported in the Statistics of U.S. Businesses (SUSB): 4.1% in 2000 and 4.27% in 2008.[18]

Industry Concentration and Firm Market Power

In this subsection, I construct the measures of industry concentration and firm market power using employment data from the LBD.

To measure the degree of industry concentration, I construct the Herfindahl-Hirschman Index (HHI) as the sum of the squared firm employment shares in industry j in year y:

$$HHI_{j,y} = \sum_f \left(\frac{emp_{f,j,y}}{emp_{j,y}}\right)^2 \tag{1}$$

where $emp_{f,j,y}$ is employment of firm f in industry j in year t. $emp_{j,y}$ is the total employment in industry j in year t. I construct two variants of the HHI measures using two- or three-digit SIC industry codes.

As the HHI's are measured using employment data for almost all private and publicly listed firms in the U.S., they can capture the degree of industry concentration more accurately than the ones constructed using publicly listed firms. Moreover, my measures of HHI cover a wider range of industries and years than the publicly available statistics reported in the Economic Census, where HHI is limited to manufacturing industries in calendar years ending in 2 or 7.[19]

I next define the market power for a firm classified into industry j as the ratio of total employment in firm f in industry j to total employment in industry j:

$$MarketPower^{E}_{f,j,y} = \frac{emp_{f,j,y}}{emp_{j,y}} \times 100 \tag{2}$$

where $emp_{f,j,y}$ is employment of firm f in industry j in year t. $emp_{j,y}$ is the total employment in industry j in year t. E represents this market power measure is constructed using employment data. I construct two variants of the firm market power measures using two or three-digit SIC codes to define industries.

A firm's market power can be interpreted as its ability in extracting rents. The measure of firm market power is constructed following the argument of Shepherd (1972) that market power is firm-specific and it depends on the firm's market share. Relative to other firms within the same market, firms with higher market share are expected to extract higher rents because they have the higher bargaining power to lower input prices, they can take advantage of economies of scale, or they have higher monopoly power in raising prices. This measure has been used by courts as a primary criterion to assess the existence of monopoly power in a specific product market because measuring market power using firm-level markup is notoriously difficult.[20,21]

Measures constructed in this section are at annual-level since the LBD reports employment as of March 12th in each year. To create a quarterly panel, I then link these measures in the year y to quarterly measures of firm workforce composition and wage patterns in the first three quarters of year y and the last quarter of year $y-1$.

Summary Statistics

Trends in Concentration and Wages: Finance vs. Non-Finance

Figure 1(a) [note: all figures and tables are at the end of the paper] plots trends of the Herfindahl-Hirschman Index(HHI) constructed by Equation (1) using 3-digit SIC. The computed HHI is averaged across industry-year cells within each of the six-year periods (the last period includes seven years, 2002-2008) using the number of employees in each cell as the weight. The average HHI can be interpreted as the degree of employer concentration the average worker faces in the financial or non-financial industries. On average, the employer concentration keeps increasing since 1990 in both financial and non-financial industries. Specifically, the average HHI concentration measure has increased by approximately 39.56% in finance, and by about 40.16% in non-finance.[22]

Figure 1(b) plots trends of average real wages computed in finance and non-financial industries. The computed average wage is averaged across firm-year-quarter cells within each of the six-year periods (the last period includes seven years, 2002-2008) using the number of employees in each cell as the weight. Figure 1(b) shows stronger growth in real wages in finance over 1990-2008: real wages on average have increased by 23.38% in finance, whereas the increase is only 8.95% in non-financial industries within the sample. By looking at changes, Figures 1(a) and 1(b) show that the increase in finance wages is around 2.6 times than the increase in non-finance wages while the changes in concentration are similar in the financial and non-financial industries. These results indicate that industry concentration creates less wage dampening effect on finance wages.

Cross-Sectional Summary Statistics

Table 1 reports summary statistics of firm-level variables from the baseline sample. Column 1 reports mean values along with standard deviation in parentheses calculated across all firm-quarter within the sample. Column (2) and (3) report mean values calculated for non-finance and financial firms respectively. The last column reports the difference between columns (3) and (2) along with statistical significance-level. Panel A reports summary statistics of firm-level wage patterns, and Panel B reports other firm characteristics including measures of firm market powers and workforce composition respectively.

Panel A shows that the quarterly average wage in finance is $10914 within the sample, which is 24% higher than the one in non-finance. Moreover, the average wage of high-skilled workers is 34.16% higher in finance. Compared to existing literature, I document a lower finance excess wage due to following reasons: 1) the calculated wage is averaged across workers from the LEHD. However, the coverage of the LEHD data used in this study only extends to 31 states, and thus I may underestimate average finance wages by excluding workers working in excluded states such as New York and Connecticut where excess wages paid by financial firms are even higher (Philippon and Reshef, 2012). 2) The frequency of my baseline sample is quarterly. The excess wages paid by financial firms, which is mainly driven by bonuses (Bell and Van Reenen, 2013), may be smoothed out by taking averages across quarters. 3) Based on my firm classification rules described in section 2.3, I drop firms spanning in both financial and non-financial industries. I also drop firms which are too diversified to be classified into one single industry. Applying these filters exclude some diversified and large firms which pay higher wages (Oi and Idson, 1999). I discuss the consequences of these restrictions in Section 7.

Panel B shows that, on average, firm market power constructed by equation (2) using two-digit (three-digit) SIC industries is 0.002 percentage points (0.011 percentage points) with a standard deviation of 0.05 percentage points (0.193 percentage points) for all firms within the sample. Among financial firms, the average of their market powers in two-digit (three-digit) SIC industries is 0.003 percentage points (0.014 percentage points) which is slightly higher than the average of non-financial firms by 0.001 percentage points (0.003 percentage points).

Panel B also reports summary statistics of variables measuring firms' workforce composition, including the average of workers' education level, the average of working experience, the share of college-educated workers and the share of male workers. Financial firms on average hire a higher share of college workers and more experienced workers as compared to non-financial firms. Consistent with findings in existing literature, these results indicate finance is a high-skill industry. Interestingly, the share of male workers in financial firms is 31.39%, which is 24.74% lower than the share in non-financial firms. The fact that financial firms hire relatively less male employees on average is consistent with the labor force statistics reported by the Bureau of Labor Statistics.[23]

Empirical Analysis

Concentration and Wages

Trends presented in Figure 1 indicate that industry concentration affects finance wages differently, but the difference can be driven by other factors varying across industries. This subsection conduct a cross-sectional analysis to estimate the difference between finance and non-finance in the correlation of concentration and average wages with controlling for other observed and unobserved factors. At one extreme, firms in more concentrated industries may extract higher rents by exercising product market monopoly power, and wages should rise when firms share these rents with workers (rent-sharing effect). Alternatively, workers in a more concentrated industry may face fewer outside options in the industry. In this case, firms have labor market monopsony power to lower wages (labor monopsony effect). The relationship between industry concentration and wages depends on the relative strength of rent-sharing effect and labor monopsony effect in a given industry.

To examine the relation between industry concentration and average wages, I estimate the following cross-sectional regression:

$$logWage_{f,j,t} = \alpha_t + \gamma_1 HHI_{j,t-4} + \gamma_2 FIN_f + \gamma_3 FIN_f \times HHI_{j,t-4} + X'_{f,j,t-4}\beta + \epsilon_{f,j,t} \quad (3)$$

where $logWage_{f,j,t}$ is the log of average wages in firm f, operating within industry j, in year-quarter t. $HHI_{j,t-4}$ is the four-quarter lagged measure of concentration in industry j; FIN_f is equal to 1 if the firm is a financial firm. $X_{f,j,t-4}$ is a vector of firm-level control variables in four-quarter lags comprising the log of average worker education level, the share of male workers, the share of college workers, the log of average working experience and the log of firm age. All regressions include year-quarter fixed effect, α_t, to control for macro-level trends in affecting wages. To control for potential time-series dependence in the residuals, I cluster the standard errors at the industry-level.

Table 2 presents the results from estimating Equation (3). Column (1) shows that, on average, wages in private sectors are negatively correlated with industry concentration measured by Herfindahl- Hirschman Index (HHI) using two-digit SIC industry codes. The negative relationship is still statistically significant at the 1% level after controlling for firm workforce composition and firm age which may affect firms' productivity and thus affect wages (column (2)). While there is a negative relationship between wages and industry concentration in non-finance, column (3) shows the relationship is positive in finance and significant at the 1% level after controlling for firm workforce composition and firm age. Precisely, a one standard deviation increase in HHI is associated with 13.78% (=0.011 × (-3.917+16.44) ×100) higher average wages in finance. Column (4) shows results are robust to HHI defined using three-digit SIC industry codes.

For robustness, I follow Grullon, et al (2018) and use total number of firms in a given two-digit SIC industry as a proxy of industry concentration. An industry is more concentrated when fewer firms remain in the industry. Consistently, column (5) shows that a decrease in the number of firms is associated with lower average wages in non-finance, whereas it is associated with significantly higher average wages in finance. In sum, these results indicate industry concentration is associated with stronger rent-sharing effect in finance.

Firm Market Power and Finance Wages

3.2.1 Main Result

While industry concentration means higher market power for firms remaining in the industry on average, previous studies show that concentration disproportionately benefits larger players within industries (Shepherd, 1972; Porter, 1979; Gale, 1972). Relative to small players in a given industry, larger ones are expected to yield higher rents for the following reasons. First, larger players' products have share-based product differentiation advantage in the sense that their products are widely advertised and recognized. Second, larger players can take advantage of economies of scale to achieve a cost advantage over rivals operating at a lower rate of output. Lastly, larger players have higher bargaining power to lower input prices or setting product prices to reflect their own interests. Therefore, market power should be firm-specific and dependent on the firm's market share. Firms with higher market power can extract higher wages as compared to firms with lower market power. Using this within-industry variation in this subsection, I examine the relationship between firm-specific market power and average wages to better identify the marginal difference in the treatment of firm market power on finance wages.

To visualize the relationship between firm market power and wages, I conduct a flexible estimation of the following equation for finance and non-financial firms separately:

$$logWages_{f,j,t} = \alpha_t + \gamma_1 D^{2nd}_{f,j,t-4} + \gamma_1 D^{3rd}_{f,j,t-4} + \gamma_1 D^{4th}_{f,j,t-4} + X'_{f,j,t-4}\beta + \epsilon_{f,j,t} \quad (4)$$

where $logWages_{f,j,t}$ represents the log of average wages at firm f, operating within industry j, in year-quarter t. $D^{2nd}_{f,j,t-4}$, $D^{3rd}_{f,j,t-4}$, or $D^{4th}_{f,j,t-4}$ are equal to 1 if the firm f's market power in year-quarter $t-4$ is respectively in the second, third or fourth quartile of firm market power distribution within the sample, where firm market power is constructed by equation (2) based on two-digit SIC industry classification. $X_{f,j,t-4}$ is a vector of workforce composition variables in 4-quarters lags comprising the log of average worker education level, the share of male workers, the share of college workers, and the log of average working experience.

I plot the coefficients of $D^{2nd}_{f,j,t-4}$, $D^{3rd}_{f,j,t-4}$, and $D^{4th}_{f,j,t-4}$ from estimating equation (4) in Figure 2a.

Overall, the relationship between firm market power and wages appears to be convex and the slope is significantly steeper in finance. Specifically, workforce composition adjusted wages (wage premium) paid by financial firms with high market power (in the fourth quartile) is about 33% higher than financial firms with low market power (in the first quartile), whereas non-financial firms with high market power only pay about 3.69% higher. The high wage premium paid by financial firms with market power above the third quartile indicate that the firms with high market power drive up the average of finance wages.

I next measure the sensitivity of wages to firm market power in finance and non-finance by estimating the following equation:

$$logWages_{f,j,t} = \alpha_t + \gamma_1 MarketPower^{E}_{f,j,t-4} + \gamma_2 FIN_f + \gamma_3 FIN_f \cdot MarketPower^{E}_{f,j,t-4} + X'_{f,j,t-4}\beta + \epsilon_{f,j,t} \quad (5)$$

where $MarketPower^E_{f,j,t-4}$ is the market power of firm f, operating mainly in industry j (two- or three-digit SIC) in year-quarter $t-4$ and firm market power is defined by equation (2). The other variables are defined as the same as in equation (4). Standard errors are clustered at firm-level.

Table 3 reports results of estimating equation (5). Column (1) confirms that financial firms pay 19.8% higher on average than non-finance private sector when controlling for unobserved macro trends. I next add in my main variable of interest: $MarketPower^E$ constructed based on two-digit SIC codes, and I also controls for firm workforce compositions which may drive the heterogeneity in productivity and wages across firms.[24] Column (2) reports the results. Overall, there is a significant and positive relationship between firm market power and average wages: a one standard deviation(0.05 percentage points) increase in firm market power is associated with 0.62% ($= 0.05 \times 0.123 \times 100$) higher wages on average. In column (3), I measure to what extent the sensitivity of average wage to the change in firm market power is different in finance by including the interaction of financial firm dummy and firm market power. The estimation results show that a one standard deviation increase in firm market power is associated with 2.64% ($= 0.05 \times (0.112 + 0.415) \times 100$) higher average wages in finance, whereas the effect is only 0.56% ($= 0.05 \times 0.112 \times 100$) in non-financial industries. The difference in the effect of firm market power on finance wage is significant at the 5% level. Column (4) shows that results also hold when I include finance-by-time fixed effects to control for time-varying differences across financial and non-financial industries.

To make sure that the positive relationship between firm market power and wages is not driven by the facts that more established firms possess higher market power and established firms pay higher wages (Dunne and Roberts, 1990a; Brown and Medoff, 2003), I add the log of firm age as an additional control and report results in column (5).[25] While more established firms pay higher wages on average, the coefficients of $MarketPower^E$ and its interaction with financial firm dummy are similar the ones reported in column (3).[26]

In column (6), I replicate the specification as in column (3) but redefine $MarketPower^E$ using three-digit SIC codes to address the concern of the coarseness of defining product market using two-digit SIC codes. Results are consistent with earlier findings. Specifically, the sensitivity of wages in finance to the change in market power is about 2.4 times (=(0.0543+0.0381)/0.0381) higher than that in non-finance. This finding to some extent mirrors what I observe in Figure 1: while there is a similar increase in market power caused by concentration in finance and non-finance, the increase in finance wage is around 2.6 times higher than the increase in non-finance.

In sum, my results show that a higher firm market power is associated with significant higher wages in finance as compared to non-finance.

3.2.2 Other Measures of Wages

1. Median Wage. To make sure that the marginal difference in the treatment of firm market power for finance and non-finance wages is not solely driven by top earners at high market power financial firms being disproportionately benefited, I repeat the specification as in column (5) of Table 3 but take the logarithm of median wages as the dependent variable. Column (1) in Table 4 shows that a one standard deviation in firm market power is associated with 1.32%($= 0.05 \times (0.0686 + 0.195) \times 100$) higher median wages in finance, whereas the effect is only 0.34% ($= 0.05 \times 0.0686 \times 100$) in non-financial industries. The marginal difference in the treatment of firm market power for finance and non-finance median wages is lower than that for average wages (column (5) of Table 3), but it is still statistically and economically significant.

2. Wages Adjusted for Cost of Living. Financial firms, especially the ones with high market power, cluster in regions with high cost of living. This raises a question that whether these firms pay higher wages to compensate for high cost of living. To test this, I adjust worker wages for state-level price index using the methodology provided by the Bureau of Economic Analysis and then aggregate to firm-level by taking the average across adjusted individual wages.[27] Then I use the logarithm of the average wages adjusted for cost of living as the dependent variable and repeat the specification as in column (5) of Table 3. Column (2) of Table 4 reports the result. The result is qualitatively and quantitatively similar to what reported in column (5) of Table 3. The consistency alleviates the concern that earlier findings are driven by the heterogeneity in cost of living across regions.

3. Male and Female Wages. In this section, I investigate whether a change in firm market power disproportionately affects average male or female wages in finance. If financial firms are able to discriminate against female workers by disproportionately sharing rents to male workers, the treatment of firm market power for male wages should be significantly higher than that for female wages in finance. To investigate this, I estimate equation (5) using the log of average wages of male or female at a given firm as the dependent variable. Results are reported in column (3) and (4) of Table 4. In finance, a one standard deviation increase in firm market power is associated with 1.99% (= $0.05 \times (0.362 + 0.0354) \times 100$) increase in male wages and 2% (= $0.05 \times (0.098 + 0.302) \times 100$) increase in female wages, and the difference is not statistically significant. Therefore, the main result of this paper is not unique to a specific gender in finance.

Exploring the Mechanism

This section explores mechanisms that may explain why market power is associated with relatively higher wages in finance. In this paper, I argue that market power is associated with two competing effects: labor monopsony effect and rent-sharing effect. In finance, the rent-sharing effect dominates labor monopsony effect and the net effect is relatively stronger. To support this argument, I provide evidence on two non-mutually exclusive mechanisms: 1) financial firms with higher market power extract relatively higher rents; 2) financial firms have to give a higher proportion of rents to their workers, especially high-skill ones, due to higher worker wage bargaining power.

Evidence on Higher Rents in Finance

With higher market power, financial firms may be able to make higher rents in several ways. First, it may be relatively easier for financial firms with higher market power to increase the complexity of their products, which is positively correlated with the hidden markup in the products (Celerier and Vallee, 2017). Second, the strong belief of "too big to fail" in finance suggests that firms with higher market power are more likely to be supported by the government when they face potential failure, and thus they may be able to borrow at a much lower cost than smaller financial firms who must borrow based on their creditworthiness (e.g., Baker et al., 2009; Ahmed et al, 2015). Lastly, it may be relatively easier for financial firms to improve operational efficiency and achieve economies of scale. For example, financial firms may be able to allocate resources or apply technology more efficiently because their production process is less geographically restricted, and thus they can generate higher revenue from each unit of assets. Under these scenarios, higher market power should be associated with relatively higher profitability in finance as compared to non-finance, and financial firms pay higher because they have more profit to share. To test this mechanism, I estimate the following equation:

$$ROA_{f,j,t} = \alpha_t + \gamma_1\ MarketPower^{E}_{f,j,t-4} + \gamma_2\ FIN_f + \gamma_3\ FIN_f \cdot\ MarketPower^{E}_{f,j,t-4} + X'_{f,j,t-4}\beta + \epsilon_{f,j,t} \quad (6)$$

where $ROA_{f,j,t}$ is the return on assets of firm f operating in industry j in year-quarter t. I follow Grullon et al. (2017) to use ROA as a proxy for profitability because Barber and Lyon (1996) argue that ROA is superior to other measures of profitability in detecting abnormal operating performance. $X_{f,j,t-4}$ is a vector of firm-level control variables in 4-quarters lags comprising the log of average worker education level, the share of male workers, the share of college workers, the log of average working experience and the log of firm age. Standard errors are clustered at firm level.

To conduct the analysis, I extract a sample of publicly listed firms from the baseline sample as constructed in Section 2.2, and match these firms with their financial statement data from Compustat through a Census internal LBD-Compustat bridge.[28]

I next estimate equation (7) using the new sample to test whether a higher firm market power allows financial firms to create relatively higher profitability. Results are reported in Table 5. Consistent with my expectation, I find a positive relationship between firm profitability and firm market power in both finance and non-financial industries. This positive correlation is more pronounced in finance. Specifically, column (1) shows that a one standard deviation (0.893) increase in firm market power is associated with 0.029 (=

$0.893 \times (0.0087 + 0.0242)$) higher ROA within finance, but only 0.0078 higher ROA within non-finance and the difference between finance and non-finance is statistically significant at the 1% level. I add the log of firm age as a control in the specification as in column (2) because firms may have a better understanding of their production functions when they get more established. I continue to find 0.024 higher in ROA within finance, and the difference in the effects of market power on ROA between finance and non-finance is statistically significant at the 5% level.

To further investigate the sources of the additional profitability in finance, I follow Grullon et al. (2017) to decompose ROA into two components: the Lerner Index and the Asset Utilization ratio. Following Aghion et al. (2005), the Lerner Index is defined as operating income after depreciation scaled by total sales. Depreciation is excluded from operating income to take into account the cost of physical capital (Hall and Jorgenson, 1967). This index approximates the extent to which prices exceed marginal costs (price-cost margins). The Asset Utilization ratio is defined as the ratio of total sales to total assets, which measures the firms' efficiency in utilizing assets to generate sales. Then I re-estimate equation (6) using the Lerner Index and the Asset Utilization ratio as dependent variables. Table 6 reports results. Overall, I find firms with higher market power are associated with higher Lerner Index in non-finance, and the positive relationship is significantly stronger in finance. Meanwhile, I find a positive correlation between market power and Asset Utilization ratio on average. However, the positive effect of market power on the Asset Utilization ratio is not significantly higher in finance as compared to non-finance in all specifications.

In sum, results from this section support the hypothesis that financial firms with higher market power are associated with relatively higher profitability due to higher price-cost margins, and thus financial firms have more rents to share with their workers relative to non-financial firms.

Evidence on Higher Bargaining Power in Finance

Financial firms may have to give a relatively higher share of rents to employees because their employees have relatively higher wage bargaining power. This mechanism should apply primarily to high-skilled workers for the following reasons: first, a high-skilled worker in finance is more likely to be matched to a larger project than in other industry (Celerier and Vallee, 2017). The high scalability makes it crucial that financial workers take sufficient care of their work and get paid more.[29] It is plausible to expect financial firms to give a relatively higher share of their rents to high-skilled employees to compensate for the high scalability, and thus wages of high-skilled workers should be more sensitive to the change in market power within finance.

Second, relative to non-finance workers, finance workers are closer to the final products such that their performance are directly linked to employers' performance. Due to this reason, the costs for rivals to evaluate workers are relatively lower and the probability of workers being poached by rivals is relatively higher in finance. Financial firms may need to share a higher fraction of rents with their workers to retain workers, and thus wages in finance should be more sensitive to the change in market power. Finance high-skill workers have even higher external visibility relative to high-skill workers in non-finance. For example, *Institutional Investor* conducts an annual poll among financial firms and reports the ranking of top money managers, analyst leaders or other roles from different financial firms.[30] Due to relatively higher scalability and external visibility, wages of finance high-skill workers should be even more sensitive to firm market power changes.[31]

To examine this mechanism, I first re-estimate equation (4) with the log of average wages of high-skilled workers as the dependent variable. As the baseline measure, high-skilled workers are individuals whose earnings are above the 90th percentile of the wage distribution in the firm-year-quarter.[32] Figure 2b plots estimation results. Compared to Figure 2a, wage premium at each quartile of firm market power over the first quartile is relatively higher in both finance and non-finance, indicating average wages of high-skilled workers are more sensitive to the change in market power. The steeper slope in finance suggests that the positive relationship between average wages of high-skill workers and firm market power is more pronounced in finance.

I then repeat the specifications as in Table 3 using the same sample with the log of average wages of high-skilled workers as the dependent variable to quantify the marginal difference in the effect of firm market power on high-skill workers' wages in finance. Results are reported in Table 7. Column (1) shows that, on average, a one standard deviation increase in firm market power is associated with 1.83% (= $0.05 \times 0.365 \times$

100) higher average wages of high-skill workers. Column (2) shows that this positive relationship is significantly stronger in finance. Specifically, a one standard deviation increase in firm market power is associated with 6.37% (= $0.05 \times (0.34 + 0.933) \times 100$) higher average wages of high-skill workers in finance, whereas the effect is only 1.7% (= $0.05 \times 0.34 \times 100$) in non-financial industries. Given the mean of wages of high-skill workers in finance is $23050 per quarter within the sample, an increase of 6.37% can be translated into a $1468.3 (= 0.0637×23050) increase in the average quarterly wage of finance high-skill workers. I find similar results in column (3) and (4) where I define firm market power using three-digit SIC.

People may question whether financial firms share a higher share of rents to high-skill workers because they have relatively higher managerial power to extract higher rents without improving firm performance. Under this hypothesis, I should observe financial firms with higher market power perform worse than non-financial firms. However, results in section 4.1 do not support this prediction because firms with higher market power are associated with relatively higher profitability within finance.

Firm Market Power and Within-Firm Inequality

Results from the last section indicate that rents are disproportionally distributed to high-skill workers. Thus, it is plausible to expect that higher market power is associated with higher within-firm inequality. In Table 8, I repeat the specifications as in Table 3 with the log difference of average top 90th percentile wages and average bottom 10th percentile wages as the dependent variable. Column (1) shows that a one standard deviation increase in market power is associated with 6.66% higher within-firm pay inequality in finance, whereas the effect is only 1.94% in non-finance. The difference between finance and non-finance is significant at the 5% level. Column (2) shows results are robust to defining industries using 3-digit SIC codes.[33]

Interestingly, the relationship between firm market power and inequality is not unique to a specific gender in both finance and non-finance. Column (3) (column (4)) in Table 8 shows the results where the 90th to 10th wage ratio of male (female) workers is the outcome variable. Higher market power is associated with higher inequality within a given gender on average, and the positive relationship is significantly stronger in finance. Even though higher inequality is found among both male and female workers at financial firms with higher market power, the effect is more pronounced on the inequality among male workers than the inequality among female workers in finance. Specifically, in finance, a one standard deviation increase in firm market power is associated with 9.35% higher pay inequality among male workers, whereas it is associated with 5.63% higher pay inequality among female workers. This result suggests that rents in finance are disproportionally distributed to high-skill male workers.

Additional Robustness Checks

Local Market Power

One alternative explanation for the difference between finance and non-finance in the sensitive of wage patterns to firm market power measured at national-level is: some non-financial firms, such as restaurants and health care services, compete locally. These firms may be able to extract high rents without a high market share in national markets. Measuring market power within national-level markets may underestimate these firms' ability in extracting rents, and thus wages paid by these firms are not sensitive to firm market power measured across a national-level market.

To test this explanation, I examine the treatment of firms' market power in local markets for their local wages and how it is different for finance wages. To construct the sample, I select workers from LEHD following the same rules discussed in Section 2.2. I then aggregate worker-firm-commuting zone level data to get wage patterns and workforce compositions at firm-commuting zone-quarter level.[34] For a firm mainly operating in industry j (two-digit SIC), its market power in commuting zone (CZ) c is defined as its employment share in industry j-CZ c:

$$MarketPower^{L}_{f,j,y} = \frac{emp_{f,c,j,y}}{emp_{c,j,y}} \times 100 \quad (7)$$

where $emp_{f,c,j,y}$ is the total employment of firm f in CZ c-industry j in year y. $emp_{c,j,y}$ is the total employment in the CZ c-industry j in year y. I link this measure in year y to quarterly measures of firm-CZ workforce composition and wage patterns in the first three quarters of year y and the last quarter of year $y-1$.[35]

I implement the following specification to estimate the marginal difference in the effect of firm local market power on local wage patterns of financial and non-financial firms:

$$y_{f,c,j,t} = \gamma_1\ MarketPower^{L}_{f,c,j,t-4} + \gamma_2\ FIN_{f} + \gamma_3\ FIN_{f} \cdot\ MarketPower^{L}_{f,c,j,t-4} + X'_{f,c,j,t-4}\beta + \alpha_t + \tau_c + \epsilon_{f,c,j,t} \quad (8)$$

where $MarketPower^{L}_{f,c,j,t-4}$ is firm f's market power in CZ c-industry j in year-quarter $t-4$. FIN_f is equal to 1 if firm f is in finance. $X_{f,c,j,t-4}$ is a vector of firm-CZ workforce composition variables in 4-quarters lags comprising the log of average worker education level, the share of male workers, the share of college workers, and the log of average working experience. α_t represents year-by-quarter fixed effects. τ_c represents CZ fixed effects. Standard errors are double clustered at firm and CZ level.

Table 9 reports results. Panel A examines average wages as the outcome variable. Column (1) of Panel A shows a one standard deviation (4.81 percentage points) higher in local market power is associated with 1.95% higher average local wages in non-finance, whereas the effect of local market power is 0.16% higher in finance. In column (2) of Panel A, I control for finance-by-CZ-by-time fixed effects to absorb unobservable industry shocks in local markets. The coefficient of interaction of FIN and $MarketPower^L$ shows that the treatment of local market power for firms' local wages is 1.7% higher in finance than that in non-finance and the difference is significant at 1% level.

Panel B repeat specifications in Panel A, but examine the average wages of high-skilled workers at firm-CZ-level as the outcome variables respectively. I continue to find more pronounced effects of firm market power on wages of high-skilled workers within finance, and the differences between finance and non-finance are significant at 1% level.

In sum, finance average wages and average wages for high-skill workers are more sensitive to the change in local market power than non-finance ones. This result alleviates the concern that the high sensitivity of finance wages is driven by firm market power defined across national market underestimating the ability of non-financial firms who compete locally to extract rents.

An additional concern of measuring firm market power at national-level is that it may underestimate firms' labor monopsony power because job search is largely local (Moretti, 2011; Molloy, et al., 2014) and the lack of labor mobility lower firms' incentives in sharing rents. The significantly positive relationship between local market power and average wages alleviates concern.

Measuring Market Power Using Sales

Using occupational data from 1990, 2000 and 2010 ACS and occupation offshorability score from Autor and Dorn (2013), I find that jobs in finance industry have the highest offshorability on average.[36] For this reason, financial firms may be able to extract higher rents by offshoring jobs and pay higher wages to domestic workers with a smaller increase in their employment share in the domestic market, which is the baseline measure of firm market power. To make sure that my results is not explained by the possibility that firm market power measured using domestic employment captures this feature in finance, I re-define firm market power as the ratio of total sales of firm f to total sales of firms in the sample classified in industry j:

$$MarketPower^{S}_{f,j,y} = \frac{sales_{f,y}}{sales_{j,y}} \times 100 \quad (9)$$

where $sales_{f,y}$ is total sales of firm f in the year y from the BR.[37] $sales_{j,y}$ is the summation of sales of firms in the sample classified in industry j in year y, where industry is defined by two- or three-digit SIC codes. S represents this market power measure is constructed using sales data. I then link this measure in the year y to quarterly measures of firm workforce composition and wage patterns in the first three quarters of year y and the last quarter of year $y-1$.

I then estimate equation (5) but replace firm market power measured using employment by $MarketPower^S$. Table 10 reports the estimation results.[38] Market power used in Panel A is calculated using two-digit SIC industries while the one used in Panel B is calculated using three-digit SIC industries. Overall, I find results which are consistent with the earlier findings: firms with higher market power measured by firm sales exhibit higher average wages and higher wages of high-skill workers, and these positive linkages are significantly stronger in financial industries. The coefficients of the interaction between finance dummy and market power in Panel A show that, within finance, a one standard deviation (0.041 percentage points) higher in market power measured by sales is associated with additional 13.08% higher average wages and 28% higher average wages of high skill workers in finance.[39]

Import Competition

Some non-financial industries may face relatively higher import exposure than financial industries, for example manufacturing and wholesale trade which take a large proportion of non-financial firms. Measuring market power using only the domestic contributions to the market may understate the degree of competition faced by firms in these industries, and this may explain why higher market power is associated with a smaller increase in rents and wags in non-finance relative to finance.

One way to examine this alternative explanation is by adjusting firm market power for import exposure. However, to my knowledge, the existing measures of import exposure are not available for non-manufacturing industries due to the complexity of measuring import value. Instead, I repeat the specification as in column (5) of Table 3 but omit firms in construction, transportation and public utilities, and services as the reference group. The intuition is that these industries should face lower import exposure as compared to other non-finance private sectors, including manufacturing, mining, and wholesale and retail trade. The alternative explanation would hold if the positive effect of firm market power on wages in finance is no longer higher than that in the new reference group. In contrast, results presented in Table 11 show that firm market power is associated with higher average wages and higher wages for high-skill workers in finance relative to those in construction, transportation and public utilities, and services. The magnitude of the difference is 2.78% ($= 0.05 \times 0.555 \times 100$) and significant at the 1% level.

Table 11 also shows that, on average, average wages and average wages for high-skill have the highest sensitivity to firm market power in finance as compared to the ones in other one-digit SIC sectors. This result highlights the distinctive role of market power in finance.

Sorting Effect Based on Worker Characteristics

The difference between finance and non-finance wages reported in Table 3 could be explained by sorting based on individual characteristics. Financial firms employing workers with higher quality would be expected to have higher market power and also pay higher wages, leading to a potential upward bias in the measured effect of firm market power in finance in cross-sectional analysis that compare different firms at a given time. To address this concern, I construct an employee-employer matched panel data set, where I can track workers across firms over time.[40] This allows me to estimate the marginal difference in the treatment of firm market power for finance and non-finance wages with controlling for time-varying and time-unvarying worker quality. Specifically, I estimate:

$$logWages\_ind_{\,i,f,y} = \phi_i + \gamma_1\ MarketPower^{\,E}_{\,f,j,y-1} + \gamma_2\ FIN_{\,f} + \gamma_3\ FIN_{\,f} \cdot MarketPower^{\,E}_{\,f,j,y-1} + X'_{\,i,f,y}\beta + \epsilon_{i,f,y} \tag{10}$$

where $logWages\_ind_{i,f,y}$ represents the log of individual i's quarterly wage at firm f in year y. ϕ_i represent individual fixed effects, which absorb the time-unvarying differences among workers in observed and unobserved characteristics. $MarketPower^E_{f,j,y-1}$ is the market power of firm f, operating mainly in industry j (two-digit SIC) in year $y-1$. FIN_f is equal to 1 if firm f is in finance. $X_{i,y}$ is a vector of time-varying controls, including year dummies interacted with education dummies, and function of worker age interacted with education dummies. Standard errors are clustered at firm level.

The results are reported in Table 12. In column (1), I only control for time fixed effect and time-variant worker characteristics. Similar to firm-level results, wages in finance are on average 18.2% higher than those in non-finance. Column (2) includes firm market power and its interaction with finance dummy as additional controls. Consistent with the firm-level cross-sectional analysis, I find higher firm market power is associated with significantly higher wages in finance than in non-finance within the individual-level sample. In column (3), I control for worker fixed effects to absorb unobserved variation in worker quality. The difference in the treatment of firm market power for individual wages stays statistically significant and positive. This result is robust to controlling for state-by-year fixed and industry-by-year fixed effects, which absorb changes in state-level policies and fluctuations in industries respectively. Documenting consistent results alleviates the concern that the marginal difference in the treatment of firm market power on finance and non-finance wages is driven by the sorting based on worker quality.

External Validity

In this section, I discuss the generalizability of the results presented in this paper. The LEHD data used in this study only cover 31 states. Given that financial firms are clustered in omitted states like New York and Connecticut in which financial firms pay even higher wage premium (Philippon and Reshef, 2012), one might question whether my results can be generalized to represent the overall finance sector. I provide two sets of evidence to mitigate this concern. First, I find similar results when I look at average per worker pay ($Wage\_lbd$) calculated using firm total payroll and total employment data from LBD (Appendix Table 18). Since the LBD covers payroll and employment of all establishments in the U.S., this measure of average wage does not suffer the problem of excluding workers who work for the same firm but located outside of the 31 states in my sample.

Second, in the unreported results, I also find higher market power is associated with significantly higher $Wage\_lbd$ within finance relative to non-finance using all firms from the LBD without controlling for firm workforce composition.[41] Moreover, I find the excess wage paid in finance within the sample of all firms from the LBD is much higher than the one I observed in my baseline sample (24%), and it is very comparable to the one documented in (Philippon and Reshef, 2012). These results suggest that excluding workers located outside of my 31 states underestimate the finance wage premium, but it does not drive the relation between firm market power and wages.

For a firm in multiple industries, I classify it to the industry in which the firm has at least 50% of its employment. This rule implicitly assumes a change in firm market power would have the same effect on the wages of workers at establishments in different industries but belonging to the same firm. To alleviate the concern that my earlier findings are driven by classifying some non-finance workers into financial industries, I replicate equation (5) using establishment-year data on average wages and industry from LBD. While I cannot control for workforce compositions due to data limitation at establishment-level, I control for firm fixed effects to absorb unobservable time-invariant firm quality. The identification is achieved using within-firm variation, comparing multiple establishments belonging to the same firm but different industries. In the untabulated results, I find the positive relationship between firm market power and average wages is still significantly stronger in finance within this establishment-year sample.

Conclusion

Increasing industry concentration in the U.S. has raised concerns that declining competition in the labor market has led to slow wage growth. While this linkage holds generally, this paper shows the finance sector has been an exception. In this paper, I explore *how* and *why* market power affects financial firms' wage-setting behavior differently from non-financial firms. Given the size of the unexplained wage premium in finance and how it has contributed to income inequality, it is critical to understand why financial firms are unique in setting wages.

Using a large sample of private and publicly listed firm data from the U.S. Census Bureau, I construct proxies for firm-specific market power to examine how the variation in firm market power explains the heterogeneity of wages within finance and how the role of firm market power within finance differs from non-finance. Overall, this paper shows that higher firm market power is associated with relatively higher wages within finance as compared to non-finance.

In this paper, I argue that rent-sharing plays an essential role in driving the more pronounced effect of firm market power on finance wages. An increase in firms' market power not only increases firms' labor market monopsony power in lowering wages by decreasing competition for hiring workers, but also increases firms' product market monopoly power by decreasing competition for selling products or buying inputs. With higher market power, firms can extract higher rents to share with their workers. As compared to non-financial firms, financial firms with higher market power pay relatively higher wages because rent-sharing effect dominates the effect of labor monopsony, and the net effect is relatively higher in finance.

I provide evidence on two non-mutually exclusive mechanisms that explain why rent-sharing is more prevalent in finance. First, I show that with higher firm market power, financial firms can extract relatively higher profits to share with workers. Market power is particularly valuable in finance than in non-finance as it allows firms to charge relatively higher price-cost margins. Second, financial firms have to give a higher fraction of rents with workers as finance workers have relatively higher wage bargaining power. This mechanism is primarily applied to high-skill workers as they are more likely to be matched with larger scale jobs in finance. Indeed, I show that wages of high-skilled workers at financial firms respond more positively to the change in firm-specific market power. As financial firms disproportionally distribute rents to high-skilled workers, I also show higher market power is associated with higher within-firm pay inequality within finance relative to non-finance.

Acknowledgments

I gratefully acknowledge helpful comments and suggestions from Paige Ouimet, David Denis, Elena Simintzi, Neville Francis, Patrick Conway, Simon Alder, Stan Rabinovich, Valentin Verdier, Huan Yang, Yue Qiu, David Leather as well as seminar participants at University of North Carolina, University of Pittsburgh, UMass Amherst, Wake Forest University, Peking University HSBC and Analysis Group for helpful suggestions. All errors are my own. I also wish to thank the participants of 2019 Society of Labor Economists Annual Conference and 2019 Labor and Employment Relations Association Annual Meeting. I thank Bert Grider and Chris Galvan for their diligent assistance with the data and clearance requests. The research in this paper was conducted while the author was a Special Sworn researcher of the U.S. Census Bureau. Research results and conclusions expressed are those of the author and do not necessarily reflect the views of the Census Bureau. This paper has been screened to ensure that no confidential data are revealed. This research uses data from the Census Bureau's Longitudinal Employer Household Dynamics Program, which was partially supported by the following National Science Foundation Grants SES-9978093, SES- 0339191 and ITR-0427889; National Institute on Aging Grant AG018854; and grants from the Alfred P. Sloan Foundation.

Endnotes

[1] This concern was raised in the Council of Economic Advisers Issue Brief (2016). The similar concern was addressed at the 2018 Jackson Hole Economic Symposium: "Within product markets, there has been a notable increase in economic activity associated with large multinational corporations along with the increased market concentration in many industries. These developments suggest that large firms today may have greater market power than in the past, and this shift may result in a decrease in competition within many industries. These shifts should concern central bankers since they likely have important linkages to observed structural changes in the global economy, including lower capital investment, a declining labor share, slow productivity growth, slow wage growth and declining dynamism." See more details at https://www.kansascityfed.org/~/media/files/publicat/newsroom/2018/pressrelease jacksonhole18.pdf

[2] Previous literature provides evidence that firms share rents with their workers. For example, Blanchflower et al. (1996) use CPS data to show that changes in wages are explained by increases in industry profitability within the manufacturing sector. Card, Deficient and Maida (2014) use employer-employee matched data from Italy to show an increase in firm-specific profitability leads to significant increases in wages. Card et al.(2018) provides a summary of recent studies on rent sharing. A string of literature provides various reasons of rent-sharing: 1. making managers' lives easier (Bertrand and Mullainatha, 2003; Cronqvist et al., 2009); 2. providing workers with incentives to keep working hard (Katz and Summer, 1989; Akerlof and Yellen, 1990); 3. poor governance (Bebchuck and Fried, 2006).

[3] This measure is used by the courts as a primary criterion to assess the existence of monopoly power in a given product market. See chapter 2 in "Competition and Monopoly: Single-Firm Conduct Under Section 2 of the Sherman Act" by Department of Justice for more details.

[4] Barber and Lyon(1996) show that ROA is a better measure than other profitability measures in detecting abnormal operating performance. Grullon et al. (2017) also uses this measure to investigate the relationship between market concentration and abnormal profits.

[5] See Abowd et al. (2006) for a more detailed description of the LEHD program and the underlying datasets that it generates.

[6] Axelson and Bond (2015) show that the financial sector is featured with high reliance on bonuses. Bell and Van Reenen (2013) use UK data to show the increase in top bankers' pay is entirely due to increased bonuses.

[7] Education is imputed for employees with missing education data (Abowd et al. 2006).

[8] 31 states include: Arkansas, Colorado, Florida, Georgia, Hawaii, Iowa, Idaho, Illinois, Indiana, Louisiana, Maryland, Maine, Minnesota, Missouri, Montana, North Carolina, New Jersey, New Mexico, Nevada, Oklahoma, Oregon, Rhode Island, South Carolina, Tennessee, Texas, Utah, Virginia, Vermont, Washington, Pennsylvania and Wisconsin.

[9] See Jarmin and Miranda (2002) for more details.

[10] Before 2002, this database was referred as the Standard Statistical Establishment List (SSEL). See DeSalvo et al. (2016) for more information about the dataset.

[11] The matching process between establishments in LBD and units in LEHD is not perfect because the LBD infrastructure is based on physical establishments while the LEHD infrastructure uses reporting units (SEIN) in a given state for a given firm. SEINs may or may not match the physical establishments identified in the LBD. Therefore, I take firm as the unit of analysis.

[12] 80% of the 1990 federal minimum wage in 2001 dollars is equal to \$1923/quarter (=0.8×\$3.8/hour ×40 hours/week × 12 weeks/quarter ×1.318).

[13] To limit the probability of data errors in the sample, I also drop all observations for individuals where wages change by extreme values in one quarter.

[14] Matching between the LBD and the LEHD is an imperfect process because the LBD infrastructure is based on physical establishments while the LEHD infrastructure uses reporting units for a given firm which are defined at state level and called SEIN. SEINs may or may not match the physical establishments identified in the LBD. For this reason, I do not conduct establishment-level analysis but firm-level analysis.

[15] Census uses SIC to define industries until 2002. Starting 1997, Census defines industries using the North American Industry Classification System (NAICS). To get SIC codes for post-2001 observations, I follow Babina (2016) to build a crosswalk between SIC and NAICS codes using LBD data between 1997 and 2001. However, the mapping between SIC and NAICS may not be a one-to-one mapping. Under the case of one NAICS code being matched with multiple SIC codes, I assign the NAICS code with the SIC code used by the most number of establishments to create a one-to-one mapping between NAICS and SIC. For example, NAICS 0001 is matched with SIC 111 for 100 establishments, but it is only matched with SIC 112 for 10 establishments. Then I assign establishments in post-2001 classified into NAICS 0001 with SIC 111. If there is a tie in the number of establishments under different SICs, then I assign the NAICS code with the SIC with the highest employment.

[16] The first two digits of SIC codes indicate the major group, and the first three digits indicate the industry group. While 3-digit SIC is a more granular way of defining product markets, it may be too narrow for some large firms whose activities span over closely related but separate markets. For example, for a large insurance company like State Farm or GEICO, they may have a similar proportion of activities within SIC 631(Life insurance), SIC 632(medical serve and health insurance), SIC 633 (fire, marine and causality insurance) and SIC 639(insurance carriers). By using a 2-digit SIC classification, I increase the probability that large corporations are grouped together as competing firms in the same industry. Grullon, Larkin, and Michaely (2017) use a similar idea when they define firm industry codes by NAICS.

[17] To reduce the probability of misclassifying workers into financial industries, I drop a small proportion of firms which have employment in both financial and non-financial industries. In the unreported results, I find baseline findings are robust to including these firms.

[18] In the SUSB, a firm is classified into 2-digit NAICS sector in which it paid the largest share of its payroll. See more details at https://www.census.gov/programs-surveys/susb/technicaldocumentation/methodology.html

[19] See more details about concentration ratios published by the U.S. Census at https://www.census.gov/econ/concentration.html

[20] See chapter 2 in "Competition and Monopoly: Single-Firm Conduct Under Section 2 of the Sherman Act" by Department of Justice for more details. Other examples mentioned in this article are: U.S. Anchor Mfg., Inc. v. Rule Indus., Inc., 7 F.3d 986, 999 (11th Cir. 1993) ("The principal measure of actual monopoly power is market share..."); Weiss v. York Hosp., 745 F.2d 786, 827 (3d Cir. 1984) ("A primary criterion used to assess the existence of monopoly power is the defendant's market share.").

[21] Measuring firm-level markup requires detailed information on product prices, quantities produced, characteristics of products and marginal costs for producing each additional unit. However, such detailed data is often not available, especially for non-manufacturing industries

[22] All observation counts and estimates are rounded according to Census disclosure policies.

[23] See more at https://www.bls.gov/cps/cps_aa2002.htm

[24] Column (2) shows financial firms pay 21.9% higher on average than non-financial firms. Compared with the result reported in column (1), the excess wage paid by financial firms is even higher after controlling for MarketPowerE and firm workforce composition. The increase in the excess wage paid by financial firms is mainly driven by the facts that financial firms have lower share of male workers and wages are positive correlated with share of male workers.

[25] Firm age is defined as the oldest establishment that the firm owns in the first year the firm is observed in the LBD (Haltiwanger, Jarmin, and Miranda, 2012).

[26] One concern of measuring firm market power at national-level is that it may underestimate firms' labor monopsony power because job search is largely local (Moretti, 2011; Molloy, et al., 2014) and the lack of labor mobility lower firms' incentives in sharing rents. Firm-local labor market-level Results presented in Section 6.1 can alleviate this concern.

[27] To adjust wages in all sample years, I use state-level price index in 2008, which is the first year the index is available. See more details about the methodology of adjusting wages for state-level price index at https://www.bea.gov/sites/default/files/methodologies/RPP2016_methodology.pdf

[28] Appendix Table 13 reports the summary statistics of key variables from the sample. Interestingly, the average wage in finance is 3.29% lower than the average wage in other industries within this sample. This wage discount may be explained by lower share of male workers at financial firms and by excluding a big proportion of private hedge funds and private equity firms which pay significantly higher wage premium (Philippon and Reshef, 2012). In unreported results, within the sample, I find financial firms on average pay higher than non-financial firms do after controlling for the share of male worker. Moreover, as public listed firms are larger on average in terms of employment, firms in this sample possess higher market power on average within their industries. Lastly, the mean of ROA in finance is 0.04, which is lower than the ROA in non-finance by 0.058. The finding that the financial sector has lower profitability is consistent with Grullon et al. (2017). As the new sample is slightly different from the sample used in my baseline analysis, I start with replicating equation (5) using the new sample to validate my earlier findings. Results are reported in Appendix Table 14. Consistent with earlier findings, I find a higher firm market power is associated with significantly higher wage in finance as compared to non-finance.

[29] For example, Kaplan and Rauh (2010) estimate that the average partner in U.S. private equity firms oversaw about $430 millions of funds in 2004.

[30] See more at ttps://www.institutionalinvestor.com/research.

[31] People may wonder whether financial firms give a higher share of rents to their workers due to higher unionization rates relative to non-financial firms. While data limitation does not allow me to provide firm-level evidence, I find unionization rates are substantially lower in finance based on statistics from the Union Membership and Coverage Database constructed using CPS. This database has been widely used by other literature on labor economics, such as Matsa (2010), Benmelech, et al. (2018). For example, 9.6% of workers in private non-farming sector are covered by labor unions in 2000, whereas only 1.8% in finance are covered.

[32] Although the education level can be a measure of skill level, it may have different meaning across different occupations or generations of workers (Phillippon and Reshef, 2012). In unreported results of robustness checks, I find similar results when defining high-skill workers as workers with at least 16 or 18 years of education, or workers with wages above the 95th or 99th percentile of the within-firm wage distribution.

[33] In unreported results, I found qualitatively similar results when I measure within-firm pay inequality by the standard deviation of wages within a given firm-year-quarter.

[34] Commuting zones are clusters of U.S. counties that are characterized by strong within-cluster and weak between-cluster commuting ties. See details about commuting zones at David Dorn's website: https://www.ddorn.net/data/Dorn_Thesis_Appendix.pdf. LEHD provides county codes for workers' employers. LBD provides county codes for firms' establishments. Commuting zones are mapped to counties through the crosswalk constructed by Dorn(2009).

[35] Appendix Table 15 reports the summary statistics of key variables in the firm-CZ-level sample.

[36] See the average offshorability of jobs in each industry in Appendix Table 16.

[37] The BR reports sales data at EIN-year level. To obtain firm-year-level sales, I match the BR with LBD on EIN-year. Some firms may have multiple EINs for tax filing purposes, and I sum up EIN-year-level sales to firm-year-level under this case.

[38] It is worth noticing that large firms generally report sales data based on their fiscal calendar such that they may not have sales data ready by the time Census collect the data. Also, some multiunit firms do not file separate tax forms for each establishment location. For these reasons, sales data for a proportion of large firms are missing (DeSalvo et al., 2016), and I have to limit the sample to a subset of firm-year-quarters from my baseline sample. Appendix Table 17 reports summary statistics of key variables from this new sample. Within this sample, I find average wages at financial firms on average are 26% higher the ones at non-financial firms between 1990-2008. The average of financial firms market powers calculated using two-digit (three-digit) SIC industries is 0.007 percentage points (0.03 percentage points) which is higher than the average of non-financial firms by 0.004 percentage points (0.008 percentage points). While this sample only includes about 60% of firm-year-quarters from the baseline sample, in unreported results, I find similar industry distributions and statistics of firm-level measures of workforce composition in these two samples.

[39] Compared to the measure of market power calculated using employment and 2-digit SIC codes, the average of market power measured using sales in finance is much higher (about 2 times) than the one in non-finance. And the standard deviation of market power in finance is smaller than the one in non-finance indicating that financial industries are more likely to be dominated by fewer firms who have high market shares. The low variation of finance market power is one potential reason to explain the large magnitudes of coefficients of the interaction between finance dummy and market power found in Panel A.

[40] To construct the sample, I apply the same filters discussed in section 2.2 to select workers from LEHD database, and also require each individual to be observed at least twice. To minimize the computing requirements of a large sample size, I only keep the wage paid in first quarter of each year for each worker.

[41] Information on workforce composition are from LEHD, so they are not available for firms which are in the LBD but cannot be matched with LEHD.

References

Abowd, Stephens, Vilhuber, Andersson, McKinney, Roemer, and Woodcock, 2006. "The LEHD infrastructure files and the creation of the quarterly workforce indicators," CES Technical Paper 2006-01.

Akerlof and Yellen, 1990," The Fair Wage-Effort Hypothesis and Unemployment," *The Quarterly Journal of Economics*. Vol. 105, No. 2, pp. 255-283.

Ahmed, Anderson, Zarutskie, 2015 "Are the Borrowing Costs of Large Financial Firms Unusual? "Finance and Economics Discussion Series 2015-024. Board of Governors of the Federal Reserve System (U.S.).

Autor, Dorn, Katz, Patterson, and Van Reenen, 2017," The Fall of the Labor Share and the Rise of Superstar Firms," Working Paper.

Autor and Dorn, 2013," The Growth of Low-Skill Service Jobs and the Polarization of the US Labor Market," *American Economic Review*, 103(5), pp. 1559-97.

Axleson and Bond, 2015," Wall Street Occupations," *The Journal of Finance, Vol. LXX(5)*, pp. 1949-1996.

Babina, Tetyana, 2016," Destructive Creation at Work: How Financial Distress Spurs Entrepreneurship." Columbia Business School Working Paper, November 21, 2016.

Barber, Brad M., and John D. Lyon, 1996," Detecting abnormal operating performance: The empirical power and specification of test statistics," *Journal of Financial Economics* 41(3), pp. 359-399.

Baker and McArthur, 2009," The Value of the "Too Big to Fail "Big Bank Subsidy," CEPR Research Paper.

Barkai, 2016," Declining Labor and Capital Shares," *Stigler Center for the Study of the Economy and the State, New Working Paper Series Working Paper.*

Bebchuk and Fried, 2006, Pay without Performance: The Unfilled Promise of Executive Compensation, Harvard University Press.

Bell and Van Reenen, 2013," Extreme Wage Inequality: Pay at the Very Top," The American Economic Review, Vol. 103, pp. 153-157.

Benmelech, Bergman and Kim, 2018," Strong Employers and Weak Employees: How Does Employer Concentration Affect Wages," Working Paper.

Bertrand and Mullainathan, 2003," Enjoying the Quiet Life? Corporate Governance and Managerial Preferences," *Journal of Political Economy,* Vol. 111(5), pp 1043-075.

Bivens, Mishel and Schmitt, 2018," It's Not Just Monopoly and Monopsony," *Economic Policy Institute Research Paper.*

Blanchflower, Oswald, and Sanfey, 1996, "Wages, Profits, and Rent-sharing," *The Quarterly Journal of Economics*, Vol. 111(1), pp 227-251.

Boar and Midrigan, 2019," Markups and Inequality," Working Paper.

Boustanifar, Grant and Reshef, 2018. "Wages and Human Capital in Finance: International Evidence, 1970-2011," *Review of Finance*, pp 699-745.

Bohm, Michael J. and Metzger, Daniel and Stromberg, Per," Since You're So Rich, You Must Be Really Smart: Talent and the Finance Wage Premium," European Corporate Governance Institute (ECGI) - Finance Working Paper No. 553/2018.

Brown and Medoff, 2003 "Firm Age and Wages," *Journal of Labor Economics*, 21(3), pp 677-697.

Card, Heining and Kline, 2013. "Workplace Heterogeneity and the Rise of West German Wage Inequality," *The Quarterly Journal of Economics*, Vol. 128(3), pp 967-1015.

Card, Cardoso, Heining, and Kline, 2018. "Firms and Labor Market Inequality: Evidence and Some Theory," *Journal of Labor Economics*, Vol. 36 (S1), pp S13-S70.

Capon, N., Fitzsimons, G., Prince, R., 1996. "An individual level analysis of the mutual fund investment decision," *Journal of Financial Services Research* 10, pp 59-82.

Carline, 2008, "Strategic price complexity in retail financial markets," *Journal of Financial Economics,* 91, pp 278-287.

Celerier, Claire and Vallee, Boris, 2017. "Returns to Talent and the Finance Wage Premium," Working Paper

Celerier, Claire and Vallee, Boris, 2017. "Catering to Investors Through Security Design: Headline Rate and Complexity," *The Quarterly Journal of Economics*, Volume 132(3), pp 1469-1508.

Cronqvist, Heyman, Nilsson and Svaleryd and Vlachos, 2009," Do Entrenched Managers Pay Their Workers More?," *The Journal of Finance*, Vol. 64(1), pp 309-339.

De Locker and Eeckhout, 2017, "The Rise of Market Power and the Macroeconomic Implications," *Princeton University working paper.*

DeSalvo, Limehouse and Klimek, 2016, "Documenting the Business Register and Related Economic Business Data," CES Technical Paper.

Dorn, 2009, "Essays on Inequality, Spatial Interaction, and the Demand for Skills," Dissertation University of St. Gallen, no. 3613.

Dunne and Roberts, 1990a, "Plant, Firm and Industry Wage Variations," Unpublished manuscript.

Gale, 1972, "Market Share and Rate of Return," *The Review of Economics and Statistics*, Vol. 55(4), pp 412-423.

Grullon, Larkin and Michaely, 2017, "Are U.S. Industries Becoming More Concentrated?" Working Paper.

Hall, Robert E., and Dale W. Jorgenson, 1967, "Tax policy and investment behavior," *American Economic Review* 57(3), 391-414.

Jarmin and Miranda, 2002, "The Longitudinal Business Database," CES Technical Paper.

Katz and Summers, 1989, "Industry Rents: Evidence and Implications," Brookings Papers on Economic Activity (Microeconomics), 209-275.

Landon, 1970, "The Effect of Product-Market Concentration on Wage Levels: An Intra-Industry Approach," *ILR Review*, 23(2) pp. 237-247.

Lemieux, MacLeod, and Parent, "Performance Pay and Wage Inequality," *The Quarterly Journal of Economics*, 124(1) pp. 1-49.

Ma, Ouimet and Simintzi, 2018, "Mergers and Acquisitions, Technological Change and Inequality," *European Corporate Governance Institute (ECGI) - Finance Working Paper No. 485/2016*

Manning, 2011, Imperfect competition in the labor market, in *Handbook of Labor Economics*, Volume 4, Amsterdam: Elsevier-North Holland: pp. 973-1041.

Matsa, David, 2010, "Capital Structure as a Strategic: Evidence from Collective Bargaining," *Journal of Finance*, 65, 1197-1232.

Molloy, Raven, Christopher L. Smith, and Abigail Wozniak, 2014, "Declining Migration within the U.S.: The Role of the Labor Market," NBER Working Paper 20065.

Moretti, Enrico, 2011, "Local Labor Markets," in Handbook of Labor Economics, vol. 4B, David Card and Orley Ashenfelter, eds. (Amsterdam: Elsevier, 2011), pp 1237-1313.

Muller, Ouimet and Simintzi, 2017, "Within-Firm Pay Inequality," *The Review of Financial Studies,* 30 (10), pp. 3605-3635.

Philippon and Reshef, 2012, "Wages and Human Capital in the U.S. Finance Industry: 1909-2006," *The Quarterly Journal of Economics*, 127(4), pp 1551-1609.

Porter, 1979, "The Structure within Industries and Companies' Performance," *The Review of Economics and Statistics*, 61(2), pp 214-227.

Shepherd, 1972, "The Elements of Market Structure," *The Review of Economics and Statistics,* 54(1), pp. 25-37.

Smets, Frank and Rafael Wouters, 2007, "Shocks and Frictions in US Business Cycles: A Bayesian DSGE Approach," *American Economic Review* 97(3), pp. 586-606.

Stiglitz, 2017, "Inequality, Stagnation, and Market Power," Roosevelt Institute Working Paper.

FIGURE 1
Trends in Industry Concentration and Wages, 1990-2008

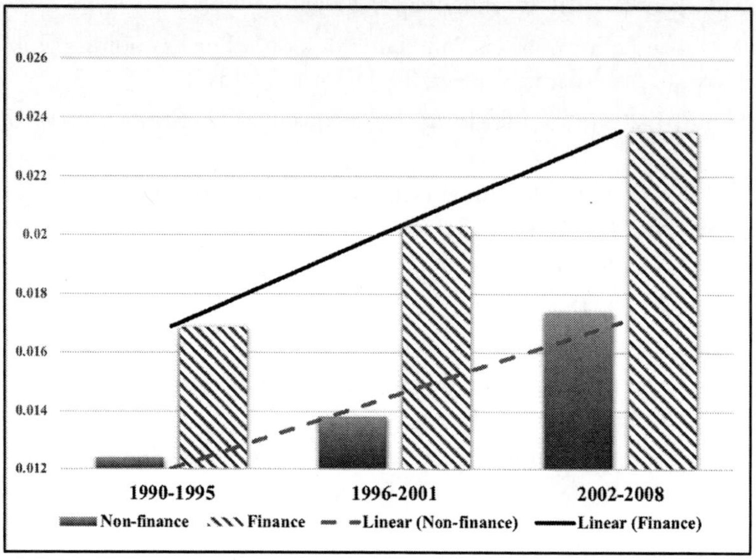

(a) Trends in Average Industry Occupation

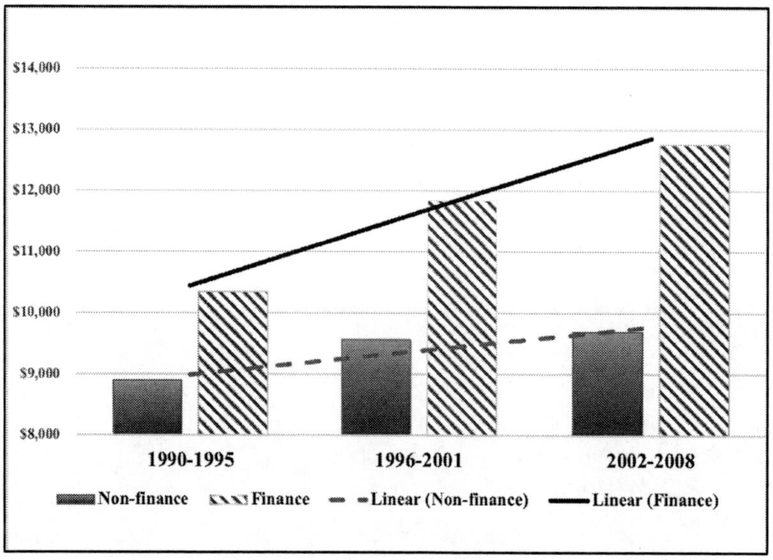

(b) Trends in Average Real Wage

Figure (a) plots trends in the employment-weighted average of the Herfindahl-Hirschman Index(HHI) constructed by Equation (1) at the 3-digit-SIC-year level by finance and non-finance sectors. Each bar represents the mean HHI in a given sector which is averaged across industry-year cells within each of the six-year periods(the last period includes seven years, 2002-2008) using the number of employees in each cell as the weight. The average HHI represents the degree of employer concentration the average worker faces in the finance or non-finance. Each straight line represent the linear trend of HHI between 1990 and 2008 in a given sector. Figure (b) plots trends in the employment-weighted average of real wages computed at the firm-year-quarter-level in finance and non-finance sectors. Each bar represents the average real wage which is averaged across firm-year-quarter cells within each of the six-year periods(the last period includes seven years, 2002-2008) using the number of employees in each cell as the weight. Each straight line represent the linear trend of average wage between 1990 and 2008 in a given sector.

FIGURE 2

Wage Patterns and Firm Market Power: Finance vs. Non-financial Industries

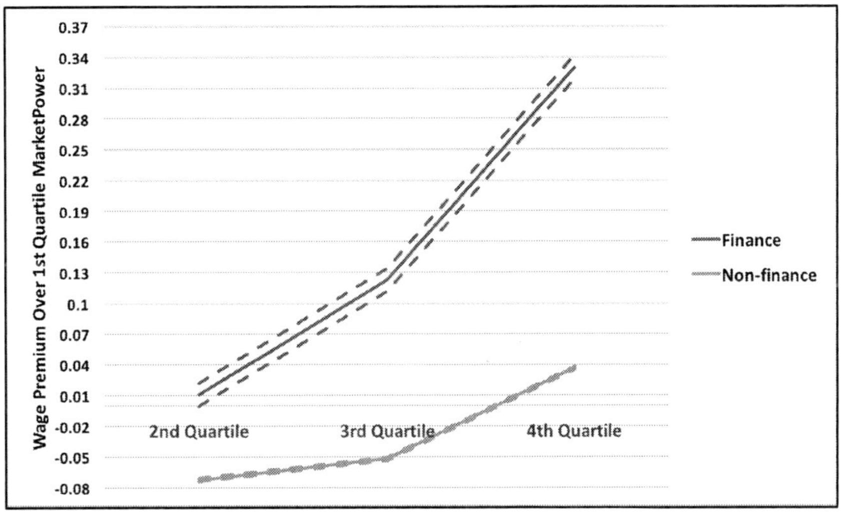

(a) Wage Premium and Market Power

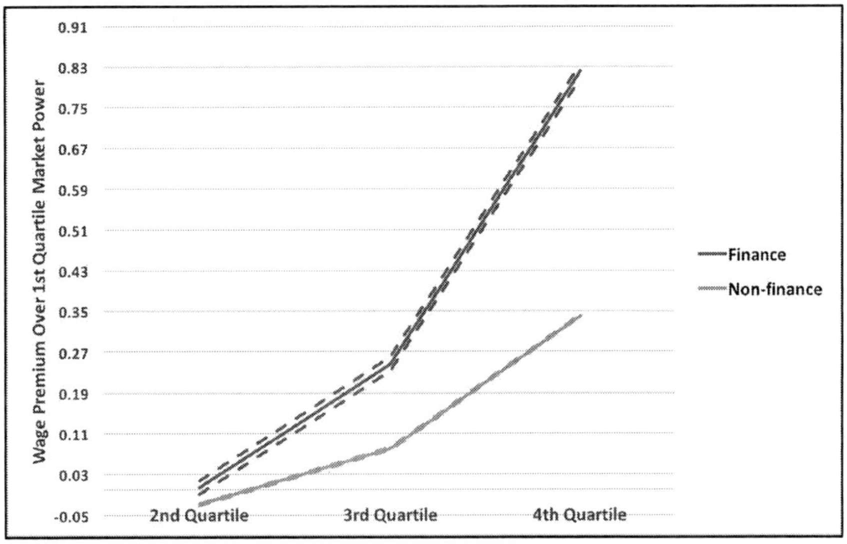

(b) Wage Premium of High-Skilled Workers and Market Power

The figures show how wage patterns in finance and non-finance respond to a change in firm market power from the first quartile of firm market power distribution within the sample. A firm's market power is measured as the firm's employment share in its industry (defined using two-digit SIC). Each figure plots regression coefficients of $D^{2nd}_{f,j,t-4}$, $D^{3rd}_{f,j,t-4}$, and $D^{4th}_{f,j,t-4}$ from equation (4), where $D^{2nd}_{f,j,t-4}$, $D^{3rd}_{f,j,t-4}$, and $D^{4th}_{f,j,t-4}$ are equal to 1 if the firm f's market power in year-quarter $t-4$ is respectively in the second, third or fourth quartile of firm market power distribution within the sample. The depended s in plot (a) and (b) are the log of quarterly average wages ($logWages$) and the log of average wages of high-skill workers ($logWages\_hskill$) respectively. The solid line indicate point estimates and the dashed line indicate 95% confidence bounds based on standard errors clustered at the firm-level.

TABLE 1

Summary Statistics: Firm Wage Pattern, Market Power and Other Characteristics

	(1) All	(2) Non-Finance	(3) Finance	(4) Difference [(3)–(2)]
Panel A: Wage Pattern				
Average quarterly wage ($)	8864 (8039)	8771 (7786)	10910 (12210)	2142***
Average quarterly wage of high-skill ($)	17430 (25650)	17180 (24630)	23050 (41880)	5869***
Quarterly wage 90th/10th percentile ratio	4.277 (5.757)	4.247 (5.704)	4.957 (6.79)	0.71***
Panel B: Firm Characteristics				
$MarketPower^E$ (2-digit SIC, %)	0.002 (0.05)	0.002 (0.051)	0.003 (0.04)	0.001***
$MarketPower^E$ (3-digit SIC, %)	0.011 (0.193)	0.011 (0.192)	0.014 (0.208)	0.002***
HHI (2-digit SIC)	0.004 (0.011)	0.004 (0.011)	0.007 (0.007)	0.003***
HHI (3-digit SIC)	0.008 (0.019)	0.008 (0.019)	0.01 (0.018)	0.003***
Average education level (year)	13.79 (1.355)	13.77 (1.354)	14.23 (1.294)	0.46***
Average working experience (year)	20.62 (6.888)	20.57 (6.886)	21.64 (6.865)	1.062***
CollegeShare (%)	35.05 (25.42)	34.72 (25.34)	42.27 (26.2)	7.546***
MaleShare (%)	55.07 (33.42)	56.14 (33.34)	31.39 (25.57)	-24.74***
Firm age	13.04 (8.545)	12.99 (8.519)	14.13 (9.02)	1.136***
Number of observations	64,790,000	62,000,000	2,795,000	

This table reports firm-level summary statistics. The sample consists of US public and private firms, and spans from Q2, 1990 through Q3, 2008. *All* refers to all observations in the sample. *Non-finance* refers to observations in financial industries. *Finance* refers to observations in non-financial industries. In columns (1) to (3) sample means (standard deviations) are computed across all-firm-quarter observations in each category. Column (4) provides differences between means in column (3) and column (2). Stars in the column (4) represent the level of p-values of testing the difference between columns 2 and 3: *** indicates $p<0.01$, ** indicates $p<0.05$, and * indicates $p<0.1$. All definitions are provided in Appendix 1. The number of observations is rounded following the Census Bureau's disclosure rules.

TABLE 2
Industry Concentration and Firm Wages

	(1) logWages	(2) logWages	(3) logWages	(4) logWages	(5) logWages
FIN			0.114***	0.186***	1.347**
			(0.002)	(0.002)	(0.642)
HHI (2-digit SIC)	–3.926***	–3.374***	–3.917***		
	(0.027)	(0.024)	(0.026)		
FIN×HHI (2-digit SIC)			16.44***		
			(0.230)		
HHI (3-digit SIC)				–0.963***	
				(0.014)	
FIN×HHI (3-digit SIC)				3.315***	
				(0.074)	
LogFirmN					0.0309
					(0.036)
FIN×LogFirmN					–0.101*
					(0.057)
Number of observations	64,790,000	64,790,000	64,790,000	64,790,000	64,790,000
R-squared	0.014	0.152	0.161	0.155	0.157
Year×QuarterFE	Yes	Yes	Yes	Yes	Yes
Workforce composition		Yes	Yes	Yes	Yes
Firm Age		Yes	Yes	Yes	Yes

This table presents estimates of the relation between industry concentration and firm average wage. The sample consists of US public and private firms, and spans from Q1, 1990 through Q3, 2008. The dependent variable is the log-transformed average quarterly wages at the firm. Wages are in 2001 constant dollars. *HHI* represents the Herfindahl-Hirschman Index. *LogFirmN* represents the log of total number of firms in a given two-digit SIC industry. Column (2)-(5) control for the four-quarter-lag of log of firm age and firm-level measures of workforce composition, including share of male workers, log of average education level, share of college workers, and log of average worker experience. All controls are lagged by four quarters, except the indicator *FIN*. Standard Errors are clustered at industry level and reported in parentheses. *** indicates $p<0.01$, ** indicates $p<0.05$, and * indicates $p<0.1$. All definitions are provided in Appendix 1. The number of observations is rounded following the Census Bureau's disclosure rules.

TABLE 3
Firm Market Power and Wages in Finance

	(1)	(2)	(3)	(4)	(5)	(6)
Variables	logWages	logWages	logWages	logWages	logWages	logWages
FIN	0.198***	0.219***	0.218***		0.217***	0.218***
	(0.0016)	(0.0015)	(0.0016)		(0.0016)	(0.0015)
$MarketPower^E$		0.123***	0.112***	0.112***	0.0971***	
(2-digit SIC)		(0.0417)	(0.0391)	(0.039)	(0.0346)	
FIN×$MarketPower^E$			0.415**	0.424**	0.4**	
(2-digit SIC)			(0.172)	(0.175)	(0.163)	
$MarketPower^E$						0.0381***
(3-digit SIC)						(0.0049)
FIN×$MarketPower^E$						0.0543***
(3-digit SIC)						(0.0161)
Number of observations	64,790,000	64,790,000	64,790,000	64,790,000	64,790,000	64,790,000
R-squared	0.014	0.151	0.151	0.151	0.154	0.151
Year×QuarterFE	Yes	Yes	Yes	Yes	Yes	Yes
Workforce Composition		Yes	Yes	Yes	Yes	Yes
Firm Age					Yes	
FIN×Year×Quarter FE				Yes		

This table presents the estimates of the effects of firm market power measured by employment on the wages of financial and non-financial firms. The sample consists of US public and private firms, and spans from Q2, 1990 through Q3, 2008. The dependent variable is the log-transformed average quarterly wages at the firm. Wages are in 2001 constant dollars. Standard errors are clustered at firm-level and reported in parentheses. *** indicates p<0.01, ** indicates p<0.05, and * indicates p<0.1. All definitions are provided in Appendix 1. The number of observations is rounded following the Census Bureau's disclosure rules.

TABLE 4
Firm Market Power and Other Measures of Wages

	(1)	(2)	(3)	(4)
	logMedWages	logWages_adj	logWages_m	logWages_f
FIN	0.213***	0.223***	0.268***	0.148***
	(0.001)	(0.0016)	(0.0025)	(0.0016)
$MarketPower^E$ (2-digit SIC)	0.0686***	0.0996***	0.0354**	0.098***
	(0.0253)	(0.0352)	(0.0157)	(0.035)
FIN×$MarketPower^E$ (2-digit SIC)	0.195**	0.389**	0.362***	0.302**
	(0.0799)	(0.161)	(0.138)	(0.125)
Number of observations	64,790,000	64,790,000	39,990,000	39,990,000
R-squared	0.154	0.154	0.12	0.075
Year×QuarterFE	Yes	Yes	Yes	Yes
Workforce Composition	Yes	Yes	Yes	Yes
Firm Age	Yes	Yes	Yes	Yes

This table presents the estimates of the effects of firm market power measured by employment on different measures of wages of finance and non-finance firms. The dependent variables are the log of median quarterly wages at the firm in column (1), the log of average quarterly wages adjusted for cost of living in column (2), the log of average quarterly wages of male workers in column (3) and the log of average quarterly wages of female workers in column (4). Wages are in 2001 constant dollars. Standard Errors are clustered at firm-level and reported in parentheses. *** indicates p<0.01, ** indicates p<0.05, and * indicates p<0.1. All definitions are provided in Appendix 1. The number of observations is rounded following the Census Bureau's disclosure rules.

TABLE 5
Firm Market Power and Profitability in Finance

	(1) ROA	(2) ROA	(3) ROA	(4) ROA
FIN	–0.0527*** (0.0054)	–0.0591*** (0.0054)	–0.0495*** (0.0054)	–0.0564*** (0.0054)
$MarketPower^E$ (2-digit SIC)	0.0087** (0.0035)	0.0056** (0.0024)		
FIN× $MarketPower^E$ (2-digit SIC)	0.0242*** (0.008)	0.021** (0.0093)		
$MarketPower^E$ (3-digit SIC)			0.0047*** (0.0012)	0.0035*** (0.001)
FIN× $MarketPower^E$ (3-digit SIC)			0.0003 (0.002)	0.0007 (0.002)
lgFirmAge		0.064*** (0.0052)		0.063*** (0.0052)
Number of observations	91,000	91,000	91,000	91,000
R-squared	0.047	0.078	0.05	0.08
Year×Quarter FE	Yes	Yes	Yes	Yes
Workforce Composition	Yes	Yes	Yes	Yes

This table presents the estimates of the effects of firm market power measured by employment on the firms' profitability within finance and non-finance. The sample consists of US public firms, and spans from Q2, 1990 through Q4, 2005. The dependent variable is the return on asset(ROA) at the firm, where ROA is defined as the EBITDA scaled by total assets at given firm-year-quarter. Besides time fixed effects, all regressions control for the four-quarter-lag of firm-level measures of workforce composition, including the share of male workers, the log of average education level, the share of college workers, and the log of average worker experience. Column (2) and (4) also control for the four-quarter-lag of log firm age. Standard errors are clustered at firm-level and reported in parentheses. *** indicates $p<0.01$, ** indicates $p<0.05$, and * indicates $p<0.1$. All definitions are provided in Appendix 1. The number of observations is rounded following the Census Bureau's disclosure rules.

TABLE 6
Firm Market Power, Price–Cost Margins and Efficiency in Finance

	Panel A				Panel B			
	(1)	(2)	(3)	(4)	(1)	(2)	(3)	(4)
	LernerIndex	LernerIndex	LernerIndex	LernerIndex	AssetUtilization	AssetUtilization	AssetUtilization	AssetUtilization
FIN	0.250*** (0.0141)	0.235*** (0.014)	0.247*** (0.0136)	0.23*** (0.0136)	−1.184*** (0.0345)	−1.201*** (0.0347)	−1.183*** (0.0339)	−1.202*** (0.0341)
$MarketPower^E$ (2-digit SIC)	0.0113** (0.0051)	0.0035 (0.0024)			0.0347** (0.0165)	0.0262 (0.0161)		
FIN× $MarketPower^E$ (2-digit SIC)	0.0653* (0.0398)	0.0574* (0.0429)			0.0604 (0.0454)	0.0519 (0.0447)		
$MarketPower^E$ (3-digit SIC)			0.0061*** (0.0015)	0.0029*** (0.0009)			0.0037 (0.0053)	0.0002 (0.0051)
FIN× $MarketPower^E$ (3-digit SIC)			0.0261*** (0.0049)	0.0273*** (0.0043)			0.0043 (0.0087)	0.0056 (0.0088)
lgFirmAge		0.158*** (0.0163)		0.156*** (0.0163)		0.170*** (0.0253)		0.174*** (0.0255)
Number of observations	91,000	91,000	91,000	91,000	91,000	91,000	91,000	91,000
R-squared	0.065	0.097	0.068	0.1	0.204	0.211	0.203	0.21
Year×Quarter FE	Yes	Yes	Yes	Yes	Yes	Yes	Yes	Yes
Workforce Composition	Yes	Yes	Yes	Yes	Yes	Yes	Yes	Yes

This table presents the estimates of the effects of firm market power measured by employment on price-cost margins and efficiency of firms within finance and non-finance. The sample consists of US public firms, and spans from Q2, 1990 through Q4, 2005. The dependent in Panel A is the *Lerner Index*, which is defined as the operating income after depreciation scaled by total sales at a given firm-year-quarter. The dependent in Panel B is the *Asset Utilization Ratio*, which is defined as total sales by total assets at a given firm-year-quarter. Besides time fixed effects, all regressions control for the four-quarter-lag of firm-level measures of workforce composition, including the share of male workers, the log of average education level, the share of college workers, and the log of average worker experience. Column (2) and (4) also control for the four-quarter-lag of log firm age. Standard errors are clustered at firm-level and reported in parentheses. *** indicates $p<0.01$, ** indicates $p<0.05$, and * indicates $p<0.1$. All definitions are provided in Appendix 1. The number of observations is rounded following the Census Bureau's disclosure rules.

TABLE 7
Firm Market Power and Wages of High-Skill Workers in Finance

	(1) logWages_hskil	(2) logWages_hskil	(3) logWages_hskil	(4) logWages_hskil
FIN	0.242*** (0.0023)	0.239*** (0.0026)	0.242*** (0.0023)	0.241*** (0.0023)
$MarketPower^E$ (2-digit SIC)	0.365*** (0.116)	0.34*** (0.111)		
FIN×$MarketPower^E$ (2-digit SIC)		0.933** (0.428)		
$MarketPower^E$ (3-digit SIC)			0.124*** (0.0156)	0.119*** (0.0158)
FIN×$MarketPower^E$ (3-digit SIC)				0.0919** (0.0401)
Number of observations	64,790,000	64,790,000	64,790,000	64,790,000
R-squared	0.093	0.093	0.093	0.093
Year×QuarterFE	Yes	Yes	Yes	Yes
Workforce composition	Yes	Yes	Yes	Yes

This table presents the estimates of the effects of firm market power measured by employment on the wages of high-skill workers at finance and non-finance firms. The sample consists of US public and private firms, and spans from Q2, 1990 through Q3, 2008. The dependent variable is the log-transformed average quarterly wages of high-skill workers at the firm. Wages are in 2001 constant dollars. Standard errors are clustered at firm-level and reported in parentheses. *** indicates p<0.01, ** indicates p<0.05, and * indicates p<0.1. All definitions are provided in Appendix 1. The number of observations is rounded following the Census Bureau's disclosure rules.

TABLE 8
Firm Market Power and Within-Firm Inequality in Finance

	(1) logWages90th_10th	(2) logWages90th_10th	(3) logWages90th_10th_m	(4) logWages90th_10th_f
FIN	0.0586*** (0.0026)	0.0604*** (0.0024)	0.116*** (0.004)	0.00292 (0.0024)
$MarketPower^E$ (2-digit SIC)	0.388*** (0.124)		0.424*** (0.134)	0.391*** (0.128)
FIN×$MarketPower^E$ (2-digit SIC)	0.944** (0.458)		1.323** (0.588)	0.647* (0.351)
$MarketPower^E$ (3-digit SIC)		0.132*** (0.018)		
FIN×$MarketPower^E$ (3-digit SIC)		0.075* (0.0433)		
Number of observations	64,790,000	64,790,000	39,990,000	39,990,000
R-squared	0.024	0.024	0.132	0.162
Year×QuarterFE	Yes	Yes	Yes	Yes
Workforce composition	Yes	Yes	Yes	Yes

This table presents the estimates of the effects of firm market power measured by employment on the wage disparities within finance and non-finance firms. The sample consists of US public and private firms, and spans from Q2, 1990 through Q3, 2008. The dependent in column (1) and (2) is the log difference of the average quarterly wages above the 90th percentile and below the 10th percentile of the quarterly wage distribution in that firm-year-quarter. The dependent s in column (3) (column (4))are the log difference of the average male (female) wages above the 90th percentile and below the 10th percentile of the male (female) wage distribution in that firm-year-quarter. Standard errors are clustered at firm-level and reported in parentheses. *** indicates p<0.01, ** indicates p<0.05, and * indicates p<0.1. All definitions are provided in Appendix 1. The number of observations is rounded following the Census Bureau's disclosure rules.

TABLE 9
Firm Local Market Power and Wage Patterns in Finance

	Panel A		Panel B	
	(1) logWages_cz	(2) logWages_cz	(1) logWages_hskill_cz	(2) logWages_hskill_cz
FIN	0.219*** (0.0014)		0.230*** (0.0022)	
$MarketPower^L$	0.00406*** (0.000)	0.00398*** (0.000)	0.0114*** (0.0001)	0.0114*** (0.0001)
FIN×$MarketPower^L$	0.00034 (0.0002)	0.00353*** (0.0003)	0.00476*** (0.0005)	0.0091*** (0.0007)
Observations	69,270,000	69,270,000	69,270,000	69,270,000
R-squared	0.194	0.197	0.136	0.138
Year×QuarterFE	Yes		Yes	
CZ FE	Yes		Yes	
Workforce Composition	Yes	Yes	Yes	Yes
FIN×CZ×Year×Quarter		Yes		Yes

This table presents the estimates of the effects of firm local market power on local wage patterns at finance and non-finance firms. The sample consists of US public and private firms, and spans from Q2, 1990 through Q3, 2008. The dependent variables are the log-transformed average quarterly wages at firm-commuting zone level in Panel A, and the log-transformed average quarterly wages of high-skill workers at firm-commuting zone level in Panel B. Wages are in 2001 constant dollars. All regressions control for the four-quarter-lag of firm-commuting zone level measures of workforce composition, including the share of male workers, the log of average education level, the share of college workers, and the log of average worker experience. Standard errors are double clustered at firm and commuting zone and reported in parentheses. *** indicates $p<0.01$, ** indicates $p<0.05$, and * indicates $p<0.1$. All definitions are provided in Appendix 1. The number of observations is rounded following the Census Bureau's disclosure rules.

TABLE 10
Firm Market Power Measured by Sales and Wage Patterns in Finance

	Panel A		Panel B	
	(1)	(2)	(1)	(2)
	logWages	logWages_hskil	logWages	logWages_hskil
FIN	0.193*** (0.0021)	0.16*** (0.0033)	0.215*** (0.0021)	0.208*** (0.0034)
$MarketPower^S$ (2-digit SIC)	0.337*** (0.0568)	0.82*** (0.143)		
FIN×$MarketPower^S$ (2-digit SIC)	3.19*** (0.146)	6.819*** (0.295)		
$MarketPower^S$ (3-digit SIC)			0.0388*** (0.0031)	0.0967*** (0.0077)
FIN×$MarketPower^S$ (3-digit SIC)			0.0554* (0.0286)	0.113* (0.0617)
Number of observations	39,090,000	39,090,000	39,090,000	39,090,000
R-squared	0.149	0.091	0.148	0.089
Year×Quarter FE	Yes	Yes	Yes	Yes
Workforce composition	Yes	Yes	Yes	Yes

This table presents the estimates of the effects of firm market power measured by sales on wage patterns at finance and non-finance firms. The sample consists of US public and private firms, and spans from Q2, 1990 through Q3, 2008. Industries are defined by two- and three-digit SIC in Panel A and Panel B respectively. In each panel, The dependent variable is the log-transformed average quarterly wages in column (1), the log-transformed average quarterly wages of high-skill workers in column (2), and the log difference of the average quarterly wages above the 90th percentile and below the 10th percentile of the quarterly wage distribution in column (3). Wages are in 2001 constant dollars. Besides time fixed effects, all regressions control for the four-quarter-lag of firm-level measures of workforce composition, including the share of male workers, the log of average education level, the share of college workers, and the log of average worker experience. Standard errors are clustered at firm-level and reported in parentheses. *** indicates $p<0.01$, ** indicates $p<0.05$, and * indicates $p<0.1$. All definitions are provided in Appendix 1. The number of observations is rounded following the Census Bureau's disclosure rules.

TABLE 11
Finance vs. Construction, Transportation and Public Utilities, and Services

	(1) logWages	(2) logWages_hskill
Construction, Transportation, Public Utilities and Services	Omitted	Omitted
Mining	0.123*** (0.004)	0.139*** (0.006)
Manufacturing	–0.0232*** (0.001)	0.0861*** (0.002)
Wholesale	0.0776*** (0.0011)	0.159*** (0.0016)
Retail	–0.323*** (0.001)	–0.332*** (0.001)
FIN	0.0655*** (0.001)	0.0625*** (0.002)
$MarketPower^E$ (2-digit SIC)	0.0858*** (0.013)	0.331*** (0.059)
Mining×$MarketPower^E$	0.0262 (0.031)	–0.131 (0.084)
Manufacturing× $MarketPower^E$	0.163*** (0.042)	0.388*** (0.135)
Wholesale×$MarketPower^E$	0.093 (0.088)	0.369 (0.354)
Retail×$MarketPower^E$	–0.0372 (0.037)	–0.189* (0.109)
FIN×$MarketPower^E$	0.555*** (0.199)	1.137** (0.472)
Observations	64,790,000	64,790,000
R-squared	0.21	0.143
Year×Quarter FE	Yes	Yes
Workforce Composition	Yes	Yes
Firm Age	Yes	Yes

This table presents results of the effect of firm market power on average wages and high-sklll wages. The sample consists of US public and private firms, and spans from Q1, 1990 through Q3, 2008. In both columns, the reference group includes firms in Construction, Transportation and Public Utilities, and Services. Standard Errors are clustered by firm and reported in parentheses. *** indicates p<0.01, ** indicates p<0.05, and * indicates p<0.1. All definitions are provided in Appendix 1. The number of observations is rounded following the Census Bureau's disclosure rules.

TABLE 12
Individual-Level Regressions: Firm Market Power and Worker Wages

	(1)	(2)	(3)	(4)
	LogWages_wk	LogWages_wk	LogWages_wk	LogWages_wk
FIN	0.182***	0.156***	0.0823***	
	(0.0127)	(0.0072)	(0.0021)	
$MarketPower^E$ (2-digit SIC)		−0.00791***	−0.00171***	0.00258***
		(0.0026)	(0.0006)	(0.0008)
FIN×$MarketPower^E$ (2-digit SIC)		0.0945***	0.0486***	0.0255***
		(0.0208)	(0.0081)	(0.007)
Observations	466,600,000	466,600,000	466,600,000	466,600,000
R-squared	0.168	0.854	0.854	0.874
Year FE	Yes	Yes	Yes	Yes
Worker FE	No	No	Yes	Yes
Edu×Year	Yes	Yes	Yes	Yes
Age×Edu	Yes	Yes	Yes	Yes
State×Year				Yes
Industry×Year				Yes

This table presents the estimates of the effects of firm market power measured by employment on the wages of finance and non-finance firms using individual-level panel data. This sample only includes the wage paid in first quarter of each year for each worker. The dependent variable is the log of real quarterly wages. Real wages are in 2001 constant dollars. FIN is equal to 1 if the worker' employer is classified as a finance firm. $MarketPower^E$ is a measure of a firm's market power in a given 2-digit SIC industry. All regressions control for year fixed effects, year fixed effects by education and function of worker age interacted with education dummies. Standard Errors are clustered by firm and reported in parentheses. *** p<0.01, ** p<0.05, * p<0.1.

Appendix 1: Variable Definitions

Average quarterly wage is the average quarterly wage in each firm-year-quarter. Wages are adjusted to 2001 dollars. *Source: LEHD*

Average quarterly wage of high-skill is the average quarterly wage of high-skilled workers in each firm-year-quarter. High-skilled workers are defined as workers whose earnings are above the 90th percentile of the firm wage distribution in that year-quarter. Wages are adjusted to 2001 dollars. *Source: LEHD*

Quarterly wage 90th/10th percentile ratio is the ratio of the average of quarterly wages above the 90th percentile to the average of quarterly wages below the 10th percentile of the quarterly wage distribution in that firm-year-quarter. Wages are adjusted to 2001 dollars. *Source: LEHD*

logWages is the logarithm of the average quarterly wage in each firm and year-quarter. Wages are adjusted to 2001 dollars. *Source: LEHD*

logMedWages is the logarithm of the median quarterly wage in each firm and year-quarter. Wages are adjusted to 2001 dollars. *Source: LEHD*

logWages_adj is the logarithm of the average quarterly wage in each firm and year-quarter. Wages are adjusted to 2001 dollars and adjusted for state-level cost of living. *Source: LEHD*

logWages_m (logWages_f) is the logarithm of the average quarterly wage of male (female) workers in each firm and year-quarter. Wages are adjusted to 2001 dollars. *Source: LEHD*

logWages_wk is the logarithm of a worker's wage in the first quarter of a given year. Wages are adjusted to 2001 dollars. *Source: LEHD*

logWages_cz is the logarithm of the average quarterly wage in each firm-commuting zone-year-quarter. Wages are adjusted to 2001 dollars. *Source: LEHD*

logWages_hskill is the logarithm of the average quarterly wage of high-skilled workers. High-skilled workers are defined as workers whose earnings are above the 90th percentile of the firm wage distribution in that year-quarter. Wages are adjusted to 2001 dollars. *Source: LEHD*

logWages_hskill_cz is the logarithm of the average quarterly wage of high-skilled workers. High-skilled workers are defined as workers whose earnings are above the 90th percentile of the wage distribution in that firm-commuting zone-year-quarter. Wages are adjusted to 2001 dollars. *Source: LEHD*

logWages90th_10th is the log difference of the average quarterly wages above the 90th percentile and below the 10th percentile of the quarterly wage distribution in that firm-year-quarter. *Source: LEHD*

logWages90th_10th_m (logWages90th_10th_f) is the log difference of the average male wages above the 90th percentile and below the 10th percentile of the male wage distribution in that firm-year-quarter. *Source: LEHD*

logWages90th_10th_cz is the log difference of the average male wages above the 90th percentile and below the 10th percentile of the male wage distribution in that firm-commuting zone-year-quarter. *Source: LEHD*

logWages_lbd is the logarithm of the average per worker pay in each firm and year-quarter. Average per worker pay is calculated as the total payroll divided by the total employment as of March 12th in each firm-year. To create a quarterly panel, the annual measure in the year y is linked to the first three quarters of year y and the last quarter of year $y - 1$. Wages are adjusted to 2001 dollars. *Source: LBD*

Average working experience is the average of workers' working experiences in each firm-year-quarter, where working experience is defined as worker age minus year of education minus six. *Source: LEHD*

Average education level is the average of workers' education levels in a given firm-year-quarter. *Source: LEHD*

MaleShare is the share of male workers (in percentage) in each firm-year-quarter. *Source: LEHD*

CollegeShare is the share of workers (in percentage) who have at least 4-year college education in each firm-year-quarter. *Source: LEHD*

firm age is defined as the oldest establishment that the firm owns in the first year the firm is observed in the LBD (Haltiwanger, Jarmin, and Miranda, 2012). *Source: LBD*

lgFirmAge is the logarithm of firm age, where firm age is defined as the oldest establishment that the firm owns in the first year the firm is observed in the LBD (Haltiwanger, Jarmin, and Miranda, 2012). *Source: LBD*

Firm employment is the total number of workers in a given firm-year-quarter. *Source: LBD*

FIN is equal to 1 if a firm is classified as a finance firm, and equal to 0 if a firm is operating in other private non-farming industries. A firm is classified as a finance firm if 1) more than 50% of its employees working in establishments belonging to one of the financial industries, and 2) all establishments belonging to the firm are in financial industries. A firm is classified as a non-finance firm if 1) more than 50% of its employees working in establishments belonging to a private non-finance non-farming industry, and 2) none of its establishments is in financial industries. *Source: LBD*

HHI (2-digit SIC) is a measure of concentration for 2-digit SIC industry in a given year-quarter. It is the summation of the square of firm employment shares in the industry as defined by equation (1). *Source: LBD*

HHI (3-digit SIC) is a measure of concentration for 3-digit SIC industry in a given year-quarter. It is the summation of the square of firm employment shares in the industry as defined by equation (1). *Source: LBD*

LogFirmN is a measure of concentration for 2-digit SIC industry in a given year-quarter. It is the logarithm of total number of firms in a given industry-year. *Source: LBD*

$MarketPower^{E}$ *(2-digit SIC)* is a firm's employment share in its main industry in a given year-quarter where industries are defined using 2-digit SIC codes. textitSource: LBD

$MarketPower^{E}$ *(3-digit SIC)* is a firm's employment share in its main industry in a given year-quarter where industries are defined using 3-digit SIC codes. *Source: LBD*

$MarketPower^{S}$ *(2-digit SIC)* is a firm's sales share in its main industry in a given year-quarter where industries are defined using 2-digit SIC codes. *Source: BR*

$MarketPower^{S}$ *(3-digit SIC)* is a firm's sales share in its main industry in a given year-quarter where industries is defined using 3-digit SIC codes. *Source: BR*

$MarketPower^{L}$ is the firm's employment share in a given industry-commuting zone-year-quarter. *Source: LBD*

ROA is the ratio of earnings before tax, interest, depreciation, and amortization (EBTIDA) to total assets in a given firm-year-quarter. The is winsorized at the 1st and 99th percentiles of its empirical distribution. *Source: Compustat*

Lerner Index is the ratio of operating income after depreciation to total sales in a given firm-year-quarter. The is winsorized at the 1st and 99th percentiles of its empirical distribution. *Source: Compustat*

Asset utilization ratio is the ratio of total sales to total assets in a given firm-year-quarter. The is winsorized at the 1st and 99th percentiles of its empirical distribution. *Source: Compustat*

Manufacturing is equal to 1 if the firm is in SIC 20-39 and is equal to 0 for firms in other industries. *Source: LBD*

Mining is equal to 1 if the firm is in SIC 10-14 and is equal to 0 for firms in other industries. *Source: LBD*

Wholesale Trade is equal to 1 if the firm is in SIC 50-51 and is equal to 0 for firms in other industries. *Source: LBD*

Retail Trade is equal to 1 if the firm is in SIC 52-59 and is equal to 0 for firms in other industries. *Source: LBD*

Appendix 2: Additional Figures and Tables

FIGURE 3
Accessible States in LEHD

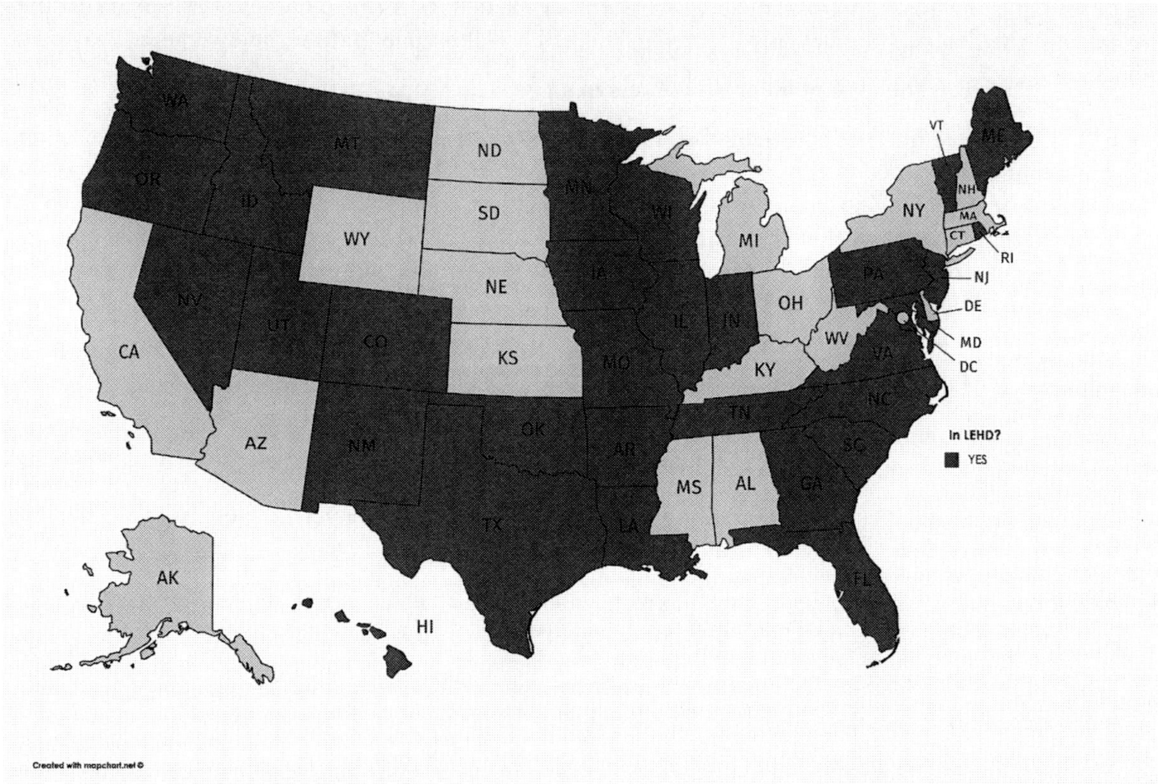

TABLE 13
Summary Statistics: Public Firm Wage Pattern, Market Power, and Other Characteristics

	(1) All	(2) Non-finance	(3) Finance	(4) Difference [(3)–(2)]
Panel A: Wage Pattern				
Average quarterly wages ($)	12780	12840	12410	–423***
	(8,542)	(8,451)	(9,077)	
Average quarterly wages of high-skill ($)	34450	33610	39600	5994***
	(35,460)	(34,510)	(40,450)	
Quarterly wage 90th/10th percentile ratio	7.358	6.953	9.848	2.895***
	(5.772)	(5.514)	(6.63)	
Panel B: Firm Characteristics				
$MarketPower^E$ (2-digit SIC)	0.219	0.239	0.093	–0.146***
	(0.893)	(0.956)	(0.258)	
$MarketPower^E$ (3-digit SIC)	0.874	0.952	0.393	–0.558***
	(2.673)	(2.788)	(1.737)	
ROA	0.089	0.097	0.04	–0.058***
	(0.173)	(0.181)	(0.093)	
Lerner Index	0.039	0.005	0.253	0.248***
	(0.420)	(0.429)	(0.278)	
Asset utilization ratio	1.241	1.408	0.217	–1.191***
	(1.026)	(0.99)	(0.525)	
Average worker age	38.9	38.78	39.67	0.9***
	(4.393)	(4.522)	(3.397)	
lgAvgEdu	14.24	14.2	14.44	0.237***
	(0.784)	(0.804)	(0.604)	
CollegeShare	42.46	41.9	45.89	3.989***
	(14.88)	(15.25)	(11.82)	
MaleShare	56.4	60.65	30.32	–30.32***
	(22.77)	(20.88)	(15.37)	
Firm age	18.53	18.23	20.41	2.185***
	(6.572)	(6.513)	(6.617)	
Number of observations	91,000	78000	13000	

This table reports firm-level summary statistics. The sample consists of publicly listed firms only, and spans from Q2, 1990 through Q4, 2005. *All* refers to all observations in the sample. *Non-finance* refers to observations in financial industries. *Finance* refers to observations in non-financial industries. In columns (1) to (3) sample means (standard deviations) are computed across all-firm-quarter observations in each category. Column (4) provides differences between means in column (3) and column (2). Stars in the column (4) represent the level of p-values of testing the difference between columns 2 and 3: *** indicates p<0.01, ** indicates p<0.05, and * indicates p<0.1. All definitions are provided in Appendix 1. The number of observations is rounded following the Census Bureau's disclosure rules.

TABLE 14
Firm Market Power and Wages in Finance (Public Firms Only)

	(1) logWages	(2) logWages	(3) logWages
FIN	0.0615*** (0.0147)	0.0394*** (0.015)	0.0502*** (0.015)
$MarketPower^E$ (2-digit SIC)		–0.0182*** (0.0046)	
FIN× $MarketPower^E$ (2-digit SIC)		0.193*** (0.0353)	
$MarketPower^E$ (3-digit SIC)			–0.0049*** (0.0015)
FIN× $MarketPower^E$ (3-digit SIC)			0.0219*** (0.0081)
Number of observations	91,000	91,000	91,000
R-squared	0.513	0.515	0.514
Year×Quarter FE	Yes	Yes	Yes
Workforce composition	Yes	Yes	Yes

This table presents the estimates of the effects of firm market power measured by employment on the wages of finance and non-finance firms. The sample consists of US public firms only, and spans from Q2, 1990 through Q4, 2005. The dependent variable is the log-transformed average quarterly wages at the firm. Wages are in 2001 constant dollars. Besides time fixed effects, all regressions control for the four-quarter-lag of firm-level measures of workforce composition, including the share of male workers, the log of average education level, the share of college workers, and the log of average worker experience. Standard errors are clustered at firm-level and reported in parentheses. *** indicates p<0.01, ** indicates p<0.05, and * indicates p<0.1. All definitions are provided in Appendix 1. The number of observations is rounded following the Census Bureau's disclosure rules.

TABLE 15
Summary Statistics: Local Market Power and Local Wage Patterns

	(1) ALL	(2) Non-Finance	(3) Finance	(4) Difference [(3)–(2)]
Average quarterly wages (CZ, $)	8839	8742	10930	2189***
	(8075)	(7788)	(12610)	
Average quarterly wages of high-skill (CZ, $)	17340	17080	23000	5929***
	(26340)	(25330)	(42180)	
Quarterly wage 90th/10th percentile ratio (CZ)	4.233	4.202	4.897	0.695***
	(5.683)	(5.629)	(6.726)	
LocalMarketPower	0.932	0.911	1.392	0.481***
	(4.810)	(4.790)	(5.193)	
Number of observations	69,270,000	66,210,000	3,061,000	

This table reports firm-commuting zone-level summary statistics. The sample consists of US public and private firms, and spans from Q2, 1990 through Q3, 2008. *All* refers to all observations in the sample. *Non-finance* refers to observations in financial industries. *Finance* refers to observations in non-financial industries. In columns (1) to (3) sample means (standard deviations) are computed across all-firm-commuting zone-quarter observations in each category. Column (4) provides differences between means in column (3) and column (2). Stars in the column (4) represent the level of p-values of testing the difference between columns 2 and 3: *** indicates $p<0.01$, ** indicates $p<0.05$, and * indicates $p<0.1$. All definitions are provided in Appendix 1. The number of observations is rounded following the Census Bureau's disclosure rules.

Table 16
Offshorability

SIC Code	Non–Farming Private Industry	Offshorability
1000–1499	Mining	–0.481
2000–3999	Manufacturing	0.227
4000–4999	Transportation and Utilities	–0.771
5000–5199	Wholesale Trade	0.395
5200–5999	Retail Trade	–0.009
6000–6799	Finance and Insurance	0.976
7000–8999	Services	0.059

This table reports employment weighted average of offshorability by industries. Higher offshorability score means jobs in the industry are more offshorable.

TABLE 17
Summary Statistics: Firm Market Power Measured by Sales and Wage Patterns

	(1) ALL	(2) Non-Finance	(3) Finance	(4) Difference [(3)-(2)]
Average quarterly wages ($)	9036	8948	11270	2326***
	(8,196)	(7,993)	(12,080)	
Average quarterly wages of high-skill($)	17920	17700	23630	5938***
	(26,060)	(25,260)	(41,150)	
Quarterly wage 90th/10th percentile ratio	4.399	4.375	5.012	0.638***
	(5.988)	(5.941)	(7.076)	
$MarketPower^S$ (2-digit SIC)	0.004	0.004	0.007	0.004***
	(0.041)	(0.041)	(0.023)	
$MarketPower^S$ (3-digit SIC)	0.023	0.022	0.03	0.008***
	(0.306)	(0.307)	(0.276)	
Number of observations	39,090,000	37,620,000	1,471,000	

This table reports firm-level summary statistics. The sample consists of US public and private firms, and spans from Q2, 1990 through Q3, 2008. *All* refers to all observations in the sample. *Non-finance* refers to observations in financial industries. *Finance* refers to observations in non-financial industries. In columns (1) to (3) sample means (standard deviations) are computed across all-firm-quarter observations in each category. Column (4) provides differences between means in column (3) and column (2). Stars in the column (4) represent the level of p-values of testing the difference between columns 2 and 3: *** indicates p<0.01, ** indicates p<0.05, and * indicates p<0.1. All definitions are provided in Appendix 1. The number of observations is rounded following the Census Bureau's disclosure rules.

TABLE 18
External Validity

	(1) logWages_lbd	(2) logWages_lbd	(3) logWages_lbd	(4) logWages_lbd
FIN	0.280*** (0.0018)	0.316*** (0.0017)	0.314*** (0.0018)	0.315*** (0.0017)
$MarketPower^E$ (2-digit SIC)			0.104*** (0.0359)	
FIN× $MarketPower^E$ (2-digit SIC)			0.461** (0.182)	
$MarketPower^E$ (3-digit SIC)				0.039*** (0.0049)
FIN× $MarketPower^E$ (3-digit SIC)				0.0669*** (0.0169)
Number of observations	64,790,000	64,790,000	64,790,000	64,790,000
R-squared	0.009	0.135	0.135	0.135
Year×Quarter FE	Yes	Yes	Yes	Yes
Workforce composition		Yes	Yes	Yes

This table presents the estimates of the effects of firm market power measured by employment on the wages of finance and non-finance firms. The sample consists of US public and private firms, and spans from Q2, 1990 through Q3, 2008. The dependent variable is the logarithm of per worker wage in the firm. Per worker wage is calculated using total pay roll divided by total firm employment, adjusted inflation to 2001 constant dollars and winsorized at 1%. Wages are in 2001 constant dollars. Standard errors are clustered at firm-level and reported in parentheses. *** indicates p<0.01, ** indicates p<0.05, and * indicates p<0.1. All definitions are provided in Appendix 1. The number of observations is rounded following the Census Bureau's disclosure rules.

Shift Attitude Changes and Employee Shift Changes Over Time

JAMES E. MARTIN
Wayne State University

We used exchange theory and employee shift changes to hypothesize attitude changes over three years. The general hypothesis is that over time, employees who moved to less day-shift work would have less positive shift satisfaction at time 2, those who moved to more day-shift work would have more positive shift satisfaction at time 2, and those who worked the same amount of day shifts over time would not differ from time 1 to time 2. The hypothesis was tested with groups who changed or did not change their amount of daytime shifts. The significant differences found supported the hypothesis.

III. LERA Best Posters, Sessions I and II

Why American Political Culture Interferes with Developing More Democratic Corporate Governance

JEROME BRAUN
Loyola University

The disinterest of America's two main political parties, the Democrats and the Republicans, for practicing populist politics is discussed as an explanation why both parties have shown little interest in supporting more democratic corporate governance. There is discussion of historical issues regarding this, and of the economic theories that justify lack of interest in pursuing a more populist politics. Discussion of the traditional paternalism of European governments is used as a partial explanation for greater formal protections for workers in Western Europe than in America.

A Non-Union Tenure System Faculty Grievance Procedure: A 45-Rear Review

ROBERT F. BANKS
JOHN L. REVITTE
Michigan State University

Introduction

This paper reviews the almost 50-year experience of handling tenure system (TS) and non-tenure system faculty and academic staff grievances at Michigan State University (MSU) with an internal non-union, but rather elaborate, dispute resolution system called the faculty grievance policy (FGP). This policy encourages dispute resolution as an initial step, with formal grievances providing for faculty hearing committees at both the initial and appeal stage, and final decisions on hearing committee recommendations by the provost or president. The review discusses the history and implementation of the FGP, the important role of the faculty grievance officer (FGO) position within the FGP who serves as grievance processor, mediator, and conciliator; reviews the outcomes of covered faculty and academic staff employment disputes within the system; and analyzes the strengths and weaknesses of the FGP.

The FGP covers many employment disputes including salary decisions and on reappointments, promotions and tenure denials for TS faculty, and similar employment status disputes. It applies to academic staff totaling some 5700 individuals including approximately 2000 tenure-system faculty. A survey of the other Big Ten Academic Alliance (BTAA) universities also is included to explore the differences and similarities between MSU's FGP experience and those of other universities. (For background information on the FGP, see https://fgo.msu.edu and http://bit.ly/2MDal4a).

The authors have interviewed almost every past living FGO and several past university presidents, provosts, members of the provost's staff, general counsel, FGP hearing panel members and councils, and members of the University Committee on Faculty Affairs (UCFA.) Surveys of FGP users about their experiences with the FGOs also have been evaluated, and FGO annual report data has been used to summarize the pattern of grievances that have occurred since 1972 to assist in an evaluation of the FGP's strengths and weaknesses.

The FGP: A Historical Context

The 1960s was a time of change and challenge for Michigan State University. A long-term president retired and was replaced by an outsider who came with new ideas.[1] It was the midst of the Vietnam War, declining state financial support, and significant personnel issues—faculty terminations processed without appeal procedures and secrecy was maintained on faculty salaries. The MSU board of trustees encouraged new policies to address these matters, including development of a non-union faculty grievance procedure.[2] A factor that encouraged this policy development was a first-time effort to unionize the faculty involving the MFT and AAUP. The voting turnout was large—82%, but only 618 faculty voted yes. Everyone interviewed—both administrators and union supporters—agree that the unionization effort, although failed, was an important impetus to the creation of a non-union faculty grievance procedure.

This paper is not to be cited, copied or reproduced without the authors' permission. Paper presented at the 71st Annual Meeting of the LERA at Cleveland, Ohio June 15, 2019

In 1970, an ad hoc committee of administrators and faculty was appointed to develop a proposal on faculty rights and responsibilities and on an FGP.[3] The initial topics were deferred so as to develop a proposal for an FGP. The document as proposed—and described in its introduction—was quite complex. The committee laid out ten major items to address but focused on key topics: whether the policy should cover procedural or substantive issues, the operational procedures for panels, and whether hearing panels or outside arbitrators should be empowered to make final decisions. The committee endorsed coverage of both substantive and procedural matters but decided that the final decision should rest with the university president. With the endorsement of academic governance, the board of trustees approved the policy in May 1972.[4] To assist in the FGP's implementation, the first two faculty grievance officials (FGOs) issued ten interpretations on procedural issues.

Steps were taken to revise the FGP in 1975, proposing largely procedural changes but suggestions that the president—as the final decision maker—could not be a grievance respondent. This issue plus faculty concerns about the narrow basis for filing a grievance, the lack of a binding arbitration option, and various perceived limitations on faculty due process rights resulted in the faculty senate voting down the proposed revision and referring it back for further study.[5] In 1978, a new proposal emerged, about which a new provost in 1979 stated several concerns in meetings of the faculty council, the most important of which was his opposition to third part arbitration, along with concerns about the broad definition of allowable grievances and procedural complexity.[6] A plan was developed to establish an administrator/faculty subcommittee to try to work out differences, but this approach produced no agreement. The provost declared an impasse and proposed a major change—that all actions dealing with tenure, reappointment, and promotion be excluded from the FGP, reserving these matters for academic administrator review. This approach also was reinforced in a proposal initiated by a new university president in 1980.[7]

At that point, deliberations on an amended grievance procedure ceased. In our view, they were superseded by three major events: deadlock between both parties; the early 1980s economic recession that resulted in major university budget reductions, program curtailments/eliminations, and faculty buy-outs under threat of termination[8]; two additional unionization campaigns in 1978 and 1981 that again registered union defeats; and the retirement of the provost and the president from their administrative roles. What remained was the original FGP, which seemed to work quite well. No further efforts to amend the policy were made for a decade, and since 1990 there have been 21 amendments of the FGP, all proposed by the FGOs and endorsed by the University Committee on Faculty Affairs(UCFA), the Office of the Provost, and the board of trustees. This close working relationship between the parties improved but did not modify the original 1972 policy. (See Appendix 1 for the most significant amendments to the FGO since the 1990s.) In addition, a detailed review of the entire FGP was conducted in 2008–09 by a group composed of the FGO and representatives from the Offices of the Provost and the General Counsel in consultation with the UCFA. The FGO policy language was edited and modified, incorporating much of the content of a previously developed FGP users' manual, but once again did not fundamentally change the 1972 version.

FGO History and Responsibilities

We will now discuss the responsibilities of the FGOs during the almost 50-year history of the FGP. One source is the professors themselves, and interviews with every FGO from 1972 through 2018, still living, were conducted. Sixteen individuals have served as FGO. Three individuals also served in acting capacities as well as in regular roles. All held senior faculty appointments at MSU, and since the early 1970s, FGOs served on a full-time basis while continuing additional academic duties. In 2012, the position was shifted to a half-time basis, although in 2019 an additional half-time appointment was being created for an associate FGO. Michael Rubner and John Revitte served continuously as FGO from 1989 to 2011, with others before and since serving shorter periods.

Since 1972, there has been some variation in FGO roles and responsibilities, but the substance of the assignment has remained consistent. Comments by FGO Rubner and Associate Provost for Academic Human Resources Robert F. Banks provide perspectives on the FGO's role. (Banks served in that capacity for 28 years.) The academic specialties of various FGOs shaped their approach to the assignment to some degree (for example, Rubner—an expert on Israeli–Palestinian relations, Revitte—labor and industrial relations, and Donohue—communications).

Rubner stated that the key FGO jobs were mediator, compromise promoter, and shuttle diplomat. Banks described these roles in more detail in 2012 remarks on the occasion of Revitte's retirement. He noted that the role involved four basic functions:

- An omnibus role—communicator, question answerer, father confessor, informal sharer of solution options
- The mediation stage, which involves meeting with both parties, playing a role as a shuttle diplomat, and a proposer of possible solutions
- The formal grievance hearing stage fully described in the FGP. The number of formal hearings is very small. Since 1972 the number of formal hearings per year numbered from zero to nine, averaging four annually, and this pales in comparison with matters addressed in informal resolutions achieved through single or multi-person conferences
- As recommender of FGP amendments and interpretations, with continuous interaction with several bodies—the provost's office, legal affairs, and the University Committee on Faculty Affairs.

Information on the FGO's role is found in annual FGO reports to academic governance. Regarding the first three roles described above, there was a large amount of activity involving the FGO office. For example, between 1984 and 2018, the office received annual inquiries ranging from 24 to 766, and the number of in-person conferences ranged annually from 106 to 185 between 1988 and 2018.

It should be noted that since the early 1970s until today, the number of tenure-system faculty at MSU has remained fairly stable at around 2000. In that same period, the number of fixed-term faculty has grown from 600 to 1300, continuing academic staff grew from 500 to 750, and fixed-term academic staff increased from 200 to 1600. Today, the total number of faculty and academic staff covered by the FGP is 5700.

In regard to the third role, the FGO schedules and is present at the hearings but is almost entirely silent. At the hearings, a faculty member not involved in the case serves as presiding officer. Also, the FGO, while a source of advice, does not have a voice or vote during the hearing panel deliberations that occur after the hearings. Thus, the FGO does not control the process in the same manner as a collective bargaining agreement arbitrator would.

In the fourth role, as recommender of FGP changes, there is evidence that FGOs took this responsibility seriously. Especially from the early 1990s until today, most FGOs have regular interactions with the provost's office and legal affairs and met monthly with the UCFA, especially with its personnel subcommittee. There also is evidence in the FGO annual reports that at least FGOs Rubner, Revitte, and Donohue were involved with the above entities in deliberations on formal changes in the FGP that were endorsed by academic governance and voted on by the eight-person MSU board of trustees selected by the citizens of Michigan for eight-year terms.

During Rubner's service as FGO, he authored an FGP users' manual, recording various UCFA actions including interpretations of the FGP; these were listed in the manual. During 2007–08, a review of the FGP involving the FGO and representatives of the provost's office and legal affairs in consultation with UCFA led to eventual academic governance endorsement and board approval in 2009 of the incorporation of many of the users' manual items as part of the FGP, as well as other changes. (See Appendix 1 for most of the important amendments to the FGP since 1990.)

Evaluation of MSU's Faculty Grievance Policy Against Estey's "Criteria"

What has been the assessment of the MSU FGP by its users and to a lesser extent by those who have to administer or deal with the policy? We use as a frame of reference an essay prepared by Marten Estey.[9] Our particular interest is on criteria for assessment of the FGP, which are briefly noted in the Estey article. We find his suggestions sensible and propose to use them. Our main focus is on the impact and success of FGP on its users, particularly faculty and academic staff.

The selected criteria are the following:

- The extent of university administrator acceptance of grievance hearing committee recommendations in grievance cases
- The cost of grievance procedures in real dollars and in time and energy
- The time spent in grievances
- The extent of faculty and academic staff use and awareness of the policy
- How often grievants win or lose

Extent of University Administrator Acceptance of Hearing Committee and Panels' Grievance Recommendations

The popular wisdom on this matter is that, whatever the hearing committee's recommendations, the university administration response tends to support the position of the provost, dean, department chairperson, or other administrators. On the other hand, administrators often believe that faculty committees are biased in favor of the grievant and thus act cautiously in response. In commentary about the MSU FGP, critics of the current system claim that its current mode of decision-making discouraged use of the system because of a lack of faith in the results. At MSU over the FGP's history, the record shows that formal grievance hearings average four per year. with annual highs of nine and a low of zero. But when the complete won/loss rate of grievants since 1972 is examined, the picture is more complex.

Data drawn from the annual reports by the FGO to MSU academic governance bodies show that in cases of formal initial hearings with panel recommendations to the provosts or other administrators and in their decisions and in presidential appeal decisions on appeal panel recommendations between 1972 and 2015, grievants fared worse by a large margin than respondents—winning in only 24.6% of the formal hearings. The presumption (such data is not available in the FGO reports) is that the provost or other relevant administrators concurred with the hearing committee recommendations supporting respondents in all those cases.

The pattern is about the same for presidential appeals during the period of 1990–2007. (Prior to 1990, no appeals are recorded, or they were combined by the FGOs with the initial hearing decisions and so they can't be separated. Additionally, since 2007, the FGO office retained no record of appeals.) In that period, there were 25 presidential appeal decisions, of which 18 were won by respondents and 7 by grievants—a 28% success rate. There is no information in these cases about the specific recommendations of the appeal hearing panels to which the president responded with a decision.

However, there is additional information obtained from some 31 provost/presidential grievance decisions collected by former FGO John Revitte covering 2003–12. For the provost, 21 case decisions were involved. In 15, the provost agreed with the hearing panel recommendations, and in all those cases, the hearing panel recommended for the respondent's position. In the remaining six with hearing committee recommendations for the grievant's position, two cases were endorsed by the provost and four were rejected. The rejections were accompanied by a rationale for the provost's response. Of the ten presidential appeals, in six the president agreed with the appeals committee, and once again those were cases in which the appeals committee agreed with the respondent. The president agreed with the appeals committee recommendations for the grievant in one case and rejected such recommendations in three cases. In the latter, those decisions were accompanied by lengthy rationales.

Again, this information covers only a short period and reflects the actions of one provost and one president. Whether the record is a reflection of results over the entire 1972–2015 period cannot be determined. Over time, grievants do fare less well than do respondents in formal decisions, but in a large fraction of the cases, the decisions of the provost/president are in accord with hearing or appeals committee recommendations that support the respondent. While the provost and president do reject some committee recommendations in cases that endorse the grievant, the numbers are a small percentage, and those decisions are accompanied by a rationale for the responses.

A hearing loss is not well received by grievants, but in a large fraction of these cases, the decision was influenced heavily by the hearing/appeal committee rather than by the responsible administrator. In almost all of these respondent recommendations endorsed by the hearing panel, support is usually unanimous. In a significant number of cases, this pattern of results shows reasonably good faith on the part of the administrator—but especially the key role played by the hearing committee in decisions supporting respondents rather the grievant.

Costs of the Grievance Policy

All policies and their implementation have costs, and one issue is whether the costs are justified by the extent of use. The direct costs include the time of the FGO and the office staff, plus supplies and services expenditures. In 2018–19, that amounted to $164,000. With no relative projection of the annual number of FGP users, it is impossible to calculate the annual cost of policies. There are other costs, largely reflected in time spent by university administrators, the FGO, individual grievants, the UCFA, hearing committee members, presiding officers, etc. For individuals involved, it is hard to determine specific costs.

In hearings, which have averaged only four per year in initial cases and much less in the fewer presidential appeals, the time of hearing panel members and presiding officers is limited—they are selected randomly and not involved in every case. Once the hearing panel members complete their FGP responsibilities, they are relieved from additional service for seven years. So, these are limited commitments for a relatively small number of individuals, and probably the time spent is in line with a faculty member's average involvement in academic governance activities at all university levels in a year.

At MSU, most grievances are settled locally and informally with involvement and support from the FGO. From 1972 to 2015, 462 grievances were settled, with 287 (62%) resolved locally. That cost is reflected in time and effort by the FGO and her/his staff. Data shows that from 1984 through 2011, inquiries to the FGO office range from 24 to 338 per year, with annual totals of more than 200 since 2005–06. Conferences on grievances, which involve more engagement, are usually attended on multiple occasions by multiple participants, as shown in data available from 2004 through 2011, with an annual total of conferences ranging from a low of 128 to a high of 273. So, while formal hearings are important, much time and effort is involved in the informal resolution of grievances, and the FGO is a key player.

How many individuals are served by the FGP? For FGO performance reviews by the UCFA, an annual survey is conducted of all who have contacted the FGO office, including those making a simple inquiry, as well as those who are actively engaged in pursuing discussions about a grievance. It is not clear whether every individual encounter is counted or only one encounter per person. In these annual surveys, the number of faculty and staff involved reported by the FGO office average about 150 individuals per year.[10] Facing concerns about general increases in university administrative costs, the Office of the Provost reduced the FGO position from full time to half time status in 2012.

A related issue is as grievances, complaints, and issues remain an ingredient of the university landscape, what are the costs of alternative arrangements for addressing these concerns? If the faculty were unionized, some costs would be borne by union supporters, but with additional cost impacts on university administrators, including costs incurred in negotiations, administration of contracts, and arbitration hearings. As we have learned in our review of other BTAA universities, there are a variety of other dispute resolution procedures: standing academic governance committees or subcommittees—usually of faculty senates—who hold hearings and make recommendations to administrators to resolve or decide grievances. Usually prior to such hearings, there is an administrative/mediation role played by a provost staff member, usually the equivalent of an associate provost for academic human resources. More infrequently, there is a grievance/dispute resolution role played by an ombudsman's office, covering faculty, academic staff, and students, usually in conjunction with the provost's office or with academic governance, and in one case, there is a procedure for an arbitrator ruling after all other steps have been completed. All these alternative complaint/dispute procedures involve time, effort, and costs that are probably not too different from those incurred by the MSU FGP. Even in non-union and no-procedure environments, there would be burdens on university administrators at all levels to address complaints, which also involve costs in time and effort. If such informal approaches did not work effectively, there would be continuing or increased concern by faculty and academic staff about a lack of adequate means to address their concerns. Such situations might result in either faculty and staff suffering in silence or resorting to litigation or resolution via other external bodies. Both results impact faculty/staff morale and productivity or involve the variable costs of court deliberations or both. Our general conclusion is that procedures addressing faculty/staff complaints vary widely in a range of environments. Despite an absence of clear information, MSU's FGP cost appears reasonable, especially when considering its success over its 50-year history.

Extent of Time Spent in Grievances Proceedings

Most of those involved, especially grievants and respondents, claim that grievance procedures take too long to be completed. Every Big Ten Academic Affiliation university has the story of one grievance or more lasting two or three plus years! MSU records are not available to make any clear time estimates. Two former long-term FGOs made a rough estimate that the average time is in the range of four months. In any case, when formal hearings are involved, there are time guidelines so that deliberations can be moved ahead, with occasional exceptions—although it should be noted that time lapses between various steps of a formal hearing are quite long.

As noted, between 1972 and 2015, 62% of grievances were settled informally. Grievances settled informally usually take a shorter time to resolve. So the time spent may be shorter in the MSU context with the high incidence of informal resolution. There are occasional but fairly few complaints about timeliness of the policy from those who participate in FGP activity, especially the grievants. Between 2009 and 2017, a question has been included in annual surveys about whether grievances were handled in a timely manner. The results show in almost in all years that 75% of responses were in the highly agree/agree categories, with only one annual result less than 70%. Continuing attention should be paid to collecting information on this point and implementing ways to reduce the length of grievance processing.

Use and Awareness of the Policy

The annual number is imperfectly projected, as about 150 individuals who interact with the FGP and FGOs appear small in comparison with the 5700 faculty and academic staff eligible to use the policy. The impression we have from this study is that there is underrepresentation. There is a small number of women and faculty and academic staff of color, as well as faculty and academic staff in general—including some 5700 individuals—eligible who seem aware of or use the policy.

Recently, the FGO office started to track users by faculty and staff assignments as well as the gender and racial and ethnic identification of FGP users. This initiative should continue in the future to keep track of this information. Attention should be given to publicizing the policy through an improved website, distribution of informational materials, and contacts with the broad potential user population by the FGO and staff. Utilization of an information specialist should be considered. The recent change of the policy's title to encourage a sense that its mission is concerned with a broader array of issues beyond grievance processing may encourage wider attention beyond traditional users. However, this initiative is just beginning, and an expansion of clientele is an issue for concern.

How Often FGP Grievants Win or Lose (from Their Perspective)

Information is provided for various aspects of interactions within the FGP. Once again, these come from annual FGO reports to academic governance for the period 1972–2015. Charts like this have been completed by FGOs each year over time. It should be noted that these categorizations are the judgment of FGOs, not grievants or respondents, that composition of these categories does not overlap, and that definitions may vary between FGOs to some degree over time. In spite of these limitations, the results show a useful pattern of results over time (see table in Appendix 3)

Of the 462 grievances addressed both formally and informally, grievants have won on 259 occasions (56% of the total). As noted, respondents prevail in formal hearings, coming out as winners in hearings (this category includes results in both initial and appeals hearings). But 216 of the 259 of grievant wins have come by way of informal resolution defined as a positive outcome without a formal hearing.

Another survey category, "grievant satisfied" (defined as a compromise satisfactory to both parties), includes 324 instances over the 1972–2015 period, or almost 29% of total interactions. So, if the complete grievance-won categories are combined with the grievance-satisfied category, it produces 511 interactions, or 45% of the total of interactions, in comparison with a total of 203 "wins" for respondents.

There is also a large number (311) recorded as "grievance dropped"—a category for which the reasons are not clearly defined. From a grievant's standpoint, the record could be better, but these outcomes could be a potential source of encouragement to use the FGP. (It should be noted that prior to this paper, these totals have never been arrayed in this fashion.)

In conclusion, a review of FGP performance in terms of its relative success, especially for grievants, gets reasonable marks for some success in grievance hearings and appeals, taking account of the relative costs of the policy and the win/loss ratio in grievance and satisfaction outcomes. Improvements in the timeliness of the policy's implementation and the achievement of more substantial participation by more (and a more diverse) faculty and academic staff with FGP eligibility are important areas for improvement.

FGO Annual Reports

The FGOs at MSU from 1972 to the present have provided annual reports to the president, provost, the Office of the General Counsel, the UCFA, and the MSU Academic Council. Annual data from these reports is summarized below (see Appendix 3 for specifics):

- Grievance cases settled (with notations of which were settled informally or via grievance hearings) and those cases still pending at the end of each academic year.
- Formal hearing settlements by subject (disputes on the amount and rationale for salary increases, reappointment, promotion and tenure decisions, plus other topics).
- Informal settlements by subject (similar to above categories).
- Total cases settled by subject and an evaluation by the FGO of whether the grievant won, lost, was satisfied with a "compromise" settlement, or dropped the case (in both informal resolution efforts and formal hearings).
- Inquiries and items included in conferences and meetings involving FGOs. These interchanges involved potential grievant faculty and academic staff and less frequently administrators, including respondent administrators. The topics from over almost 50 years were not described in the same words in these reports, but there is considerable consistency. Included were such issues as grievant rights under the FGP; retention, promotion, and tenure decisions; evaluation and salary decisions; potential discipline for violation of policies; claims of harassment by an administrator; disputes over department, school, college, and/or university bylaws; and other common faculty and academic staff subjects of concern.

Big Ten Academic Alliance Survey

Michigan State University is one of 14 universities that are full members of the Big Ten Conference. Most recognize this is an alliance of major, historically Midwest, universities that are involved in intercollegiate sports. However, the Big Ten collaborates on academic matters, and for many years, those interests were under an umbrella called the Committee on Institutional Cooperation, which today is called the Big Ten Academic Alliance (BTAA). The University of Chicago left the Big Ten in 1946 for the purpose of intercollegiate athletics but remains connected on academic topics and information sharing on non-athletic concerns.

The authors of this paper speculated for years that the MSU FGP system was unique among the Association of American Universities and Big Ten institutions in terms of having an elaborate grievance, complaint, and dispute resolution procedure for faculty and academic staff. The MSU provost and general counsel offices, sometimes along with some of the FGOs, several times tried to survey other Big Ten institutions on this topic. Historically, institutions such as the Universities of Maryland and Michigan also conducted surveys. But the information gathered was limited, and the response rates were not complete.

The authors reviewed all the BTAA websites, and it appeared that MSU was less unique than previously imagined. To more fully understand the current reality about the range of BTAA approaches on this topic, a more detailed survey would be useful. Under the auspices of the MSU Provost's Office and with support from MSU Qualtrics, a survey instrument was designed and sent out in spring 2019. The major purpose of the survey was to understand how Big Ten universities handle grievances, complaints, and disputes of their faculty and staff that are not covered by collective bargaining agreements. At present, only 9 of the 15 BTAA universities have responded, but our intent is to persist and obtain a response from them all.

The survey begins with a basic question of how many faculty and academic staff are covered by a collective bargaining agreement, as well as how many are covered but can use a separate non-union "grievance, compliant, and/or dispute resolution system." At MSU, those numbers are currently 500 and 700, respectively. About two dozen other questions follow, such as who among your faculty and academic staff can use the non-union system for dispute resolution; which administrators can have a grievance, complaint, or dispute filed against them; on what grounds may a grievance, complaint, or dispute be filed (over such issues as reappointment, promotion and/or denials, salary, etc.); timeliness deadlines and other jurisdictional requirements for a proper grievance, complaint, or dispute filing; and questions regarding both informal and formal dispute resolution steps under each university's policy for non-union dispute resolution.

Grievant and Administrator "User" Assessments of FGP

Unfortunately, there are no general historical assessments of the FGP directly as a policy. A survey beginning in 1991 evaluated only the FGO. Approximately ten years, ago a more elaborate online survey was instituted covering both the FGO and the FGO office staff, with responses broken down by type of employee who made use of the policy, including gender and ethnic/racial identity, and by participant roles in the system: (grievants, respondents, hearing panel members and presiding officers, and those making general inquires of the office). Thus, we have relied on the assessment information that is available, including inferences from data collected from surveys and from interviews with participants, largely central administration senior staff including presidents, provosts, and others.

The authors had access to a limited number of surveys on the FGO's individual job performance. Historically, such reviews were done annually by the UCFA in its role as a participant in yearly salary evaluations. This is not inappropriate because the FGO's role is the centerpiece of the FGP, but the early results do not provide a general assessment of the policy. Early in the history of the FGP, the UCFA interacted with the FGO informally, reviewed the mandated annual reports, and provided oral commentary on FGO performance to the Office of Provost. It was only in the 1990s that a system of FGO reviews was done annually, and at intervals of five-years, FGO reappointment reviews became operational.

There is limited availability of five-year reviews was because none of the FGOs appointed before 1983 served for a five-year period, and that is also the case for the four FGOs who have served since 2015. Four five-year reappointment review results should be available. FGO Mary Lee Brady's 1983–88 review has been misplaced, as was the second five-year review of FGO Michael Rubner, although the first five-year review is available. The one five-year review covering FGO John Revitte is available. Annual review forms are available since 2000, but earlier annual reviews are not available.

Initially, the five-year reviews of the FGO involved surveys of faculty who had interacted with the FGO informally or in formal grievances. Such names were picked by the FGO office, with no involvement from the FGO. The reviews were paper and pencil surveys. In the 2000s, the survey was converted to a web-based form, and the groups questioned expanded to include involved faculty and unit administrators, counsel, hearing panel chairpersons, and all university academic administrators. The number of individuals from whom a survey response was requested provides limited information on how many individuals made use of the policy. The web-based survey had higher response rates than the paper and pencil variety, although responses overwhelmingly came from faculty member (as users) in both survey types.

Records from two five-year review surveys are available (1995 and 2009). Both surveys covered a small number of the total of FGP participants, with response rates between 40 and 50%. The 1995 survey included a summary of annual reviews as well as the five-year review. Both reviews had highly positive FGO ratings at or close to a 90% rating, combining the strongly agree/agree responses with a 50% plus response rate in 9 of 13 categories. The review summary reported an overall survey response rate of 48%, with 114 of 237 surveys reported for 1994, the last year of the five-year review period.

A second report in 2009 (involving a second FGO) again provided positive ratings in both the five-year review and the summary of annual performance reviews. Question responses noting strongly agree/agree showed a 80 to 90% positive response in 12 categories focusing on such topics as accessibility, being knowledgeable about policies and procedures, and maintaining confidentiality. The more challenging categories—diversity, mediation, impartiality, and problem solving—received a somewhat lower assessment, but those ratings were above 80%. In additional written comments, respondents noted a significant positive increase in FGO performance as mediator and problem solver. Some very positive comments referred to the FGO as "a professional." Of the 172 individuals invited to complete the survey, 75 responded—a response rate of 44%. The survey was completed by 57 faculty members, 13 administrators, and 5 academic staff.

Historically, there were additional annual reviews for the FGOs, but those were not available to the authors. A more general survey on both the policy and the FGO are available for 2008 through 2017. (Because of a transition to a new FGO in 2014, there were only 19 responses, and it was decided not to include those.) The review category result reflect, in general, the outcome of the five yearly reviews noted previously. These responses were aggregated using the top two categories of strongly agree/agree as follows:

- Accessibility—above 80% with most in the 90s
- Knowledge of policies and procedures—in the high 80s
- Confidentially maintained in the process—all above 90
- sensitivity to diversity issues—mid 70s (this question was not included in some surveys).

The more challenging areas had lower scores:

- Skill as a mediator—in the 70s but some lower
- Problem solving—in the 80s but some lower
- Fairness in administration—mid 80s

In the general category rankings, satisfaction with the process was in the mid 70s, and satisfaction with the outcome ratings were occasionally in the 70s but mainly in the mid 50s.

This later data mainly reflects commentary by those who were participants in formal hearings, and such an environment always creates winners and losers. In particular, the 2015 survey results were less positive. In part, the results were related to the existence of a new and inexperienced FGO, along with some questions about (and discomfort with) the shift away from mediation and formal hearings toward a strong emphasis on improved communication and mediation as the key to issues in dispute resolution. Those themes appear frequently in open-ended comments provided in the survey. There was strong support of the FGP support staff for their steady improvement in knowledge about the policy and its implementation requirements.

Comments from the UCFA were highly positive for both FGOs, and two of the three with five years of service were recommended for reappointment. In the 1995 review summary, the UCFA commented that the FGO had "a strong record of settling grievances informally, a key role in improving and proposing changes in the FGP and in providing continuing support of the policy's integrity." The 2009 review summary noted the FGO's "strong role and his participation in the reform and improvement of the FGP."

Senior administrators were generally positive about the work of the FGOs and the operation of the FGP, as reflected in the interviews with three presidents and five provosts who served since the FGP was established. Even President Mackey and Provost Winder (both now deceased)—who in the 1970s were not supportive of the FGP as originally conceived and proposed a much narrower approach in the policy—did not raise those issues again before they stepped down from their roles as administrators. No administrators were supportive of faculty unionization and collective agreements, but they all believed that a non-union faculty grievance procedure was important for faculty and academic staff to address grievances, disputes, and complaints, especially in a large, diverse university like MSU. Many unit administrators had very little FGP involvement or experience with the FGP. Those unit administrators directly involved in grievances, especially in hearings, were concerned about the time and effort involved. Some claimed that potential involvement in grievance hearings made them overly cautious about negative personnel decisions. In the survey results noted above, there was a quite limited response by administrators as such came mainly from those participants as respondents in the small number of annual grievances.

Finally, it should be noted that since the mid 1970s, there has been no general disagreement registered with the FGP and it has a record of good performance. The UCFA has been supportive of the FGP and worked closely with the FGOs and the Office of the Provost to make incremental improvements in the policy. The press for some form of third-party arbitration and for the unionization of the faculty has generally disappeared, and their reappearance appears unlikely. Whatever imperfections and limitations of the data, the evaluations are largely positive, and the FGP retains the confidence of the faculty and academic staff, the UCFA, and the university administration.

Conclusions

The MSU FGP was born in a time of change and turmoil. A new outside president, university financial stress, festering faculty personnel issues, and efforts to unionize the faculty all encouraged its emergence. The committee selected to design the new procedure came up with a novel approach but, to date, the origin of those ideas is unclear. The basic outline for the policy included a jury-like system of initial hearing and appeal panels populated by faculty members with a focus on both procedural and substantives issues as a basis for a grievance and with the president of the university as final decision maker.

The centerpiece of the system was the appointment of an FGO with substantial authority to administer hearings and to settle most grievances informally. This position had the support of the Office of the Provost and was provided substantial powers to obtain information based on the FGO's own judgment of need and role of gatekeeper as to who has access to the procedure, at first informally and then through a system written into the procedure itself. Early FGOs played important roles as interpreters of various aspects of the policy.

A key element of the FGP that reinforces the role of the FGO is the existence of a formal hearing procedure that, if invoked, produces a decision. An individual considering formal filing of a grievance has to recognize the costs of such action: the need to prepare for a hearing, recruitment of counsel, and keeping in mind the impact that an such an action may have on relationships with departmental/school colleagues, including the unit's chair or director, which may have lasting effects. The FGO operating in this environment has leverage to encourage a grievant to consider alternative resolution techniques involving problem solving and mediation. That does not mean disregarding the issues involved in the grievance, but it does provide an opportunity to explore the cost and benefits of formal versus informal procedures and to suggest alternatives. In some cases, apart from reviewing the general context, it may involve the FGO examining the substance of the grievance and occasionally concluding there is a limited chance of winning, while being clear that the grievant can make has her/his own choice on how to proceed. As shown in this paper, a large percentage of grievances under the FGP are resolved informally, which involves either a complete or partial resolution. It also is the case that a substantial number of grievances are withdrawn without resolution at the request of the grievant based either on an assessment of the costs and benefits associated with a formal hearing or other considerations. Whatever the case, we believe that the existence of a formal hearing procedure as a part of the FGP provides the FGO important assistance with achieving its major objective: the resolution of grievances locally and informally usually a problem-solving, mediation approach.

The final element in this important triad was the UCFA, a standing committee of MSU academic governance, which was to serve as a grievance advisory body and in time became a major vehicle for making changes in the policy and its interpretation involving a close relationship and collaboration with the FGO and the Office of the Provost. The board of trustees had the authority to approve amendments, as did the major bodies of academic governance, but the source of the energy for change came from three previously mentioned entities.

The failure to reach agreement in the late 1970s–early 1980s dispute between the president and provost on the one hand and the faculty council on the other over the nature of an amended FGP strengthened the position of the original policy. Its status was also impacted by the need to focus on issues involved in a major financial crisis for the university in the early 1980s. The failure of three unionization efforts from the early 1970 to the 1980s by a two-to-one margin enhanced the policy's status as the only mechanism to protect faculty and academic staff rights. By the early 1980s, the FGP was "the only game in town." Since the 1990s, the FGP was modified in a number of important ways through joint interaction between the FGO, the

UCFA, and the Office of the Provost, which improved and rationalized the policy in many ways. These efforts, however, did not change the original character of the policy.

FGP performance was evaluated according to several criteria. In formal hearings, which averaged about four per year over the policy's history, grievants did not fare as well as respondents. They did much better when the record of grievances settled informally are included, an taking into account the results included in a separate a category titled "grievant satisfied." Forty-five percent of the total interactions recorded in the 1972–2015 period showed grievants coming out ahead. Data—although somewhat limited—shows that in many initial grievance hearings or appeals, the provost or president reached an agreement with the panel's recommendations, which involved the panel finding against the grievant. Thus, losses recorded for the grievant in those cases comes from the ruling of the faculty panel and not from the provost/president overturning a panel recommendation, although that did occur in some cases. A conclusion was reached that the costs of the policy were reasonable. The extent of resolution timeliness was unclear due to lack of reliable information and the small number of eligible individuals using the policy over time—an total of approximately 150 annual users versus some 5700 individuals using the policy appears to be a major issue.

Information on user administration assessment of the policy was examined. Although the information sources were limited because previous annual evaluations were inaccessible, the safe conclusion it that there is high satisfaction with the policy. This judgment is based on a review of several batches of annual reviews and two five-year reviews related to appointment continuations for two FGOs, both of whom were recommended for reappointment. As noted in the section on assessment of the policy, the results are generally positive in individual annual reviews and in the five yearly reviews.

The UCFA is at the centerpiece in reviewing these materials and in providing oral and written commentary to the provost. In the 1990s, a system of paper and pencil surveys was introduced and sent to all users of the policy by the FGO office staff. Sometime in the early 2000s, a web-based survey form was introduced in collaboration with the staff of the Institute for Public Policy for Social Research. Those surveys have a wider focus than on the FGO alone and also include the policy in general. In addition, the data is broken out by employee type; gender, ethnicity, and race; and the participants' roles in the grievance policy. Although the data include only a few years, it is very useful for policy evaluation.

Senior university administrators whom we interviewed are supportive of the policy. None of those administrators favored unionization of the faculty or third-party resolution of grievance outcomes. But they were supportive of a policy for non-union faculty and academic staff that provided a means to address grievances, disputes, and complaints. Many administrators lower down the administrative chain either had no or limited experience with the policy. Some of those who had such experience were concerned about the time and effort involved, and some also claimed that potential involvement in grievance hearings made them overly cautious about making negative personnel decisions.

Over time, the UCFA has been very supportive of the policy and the various FGOs. For almost 40 years, the press for faculty unionization—such as occurred recently for a small union of fixed-term faculty—has dissipated. That is also the case for the idea of binding, third-party arbitration for grievances, and reemergence of those proposals seems unlikely. Whatever its limitations and imperfections, the FGP appears to retain the confidence of the faculty and academic staff, the UCFA, and the university administration.

As of 2019, the MSU FGP retains the basic structure that was conceived in 1972 but with various amendments and interpretations to improve the policy. Not all in the university community agree that the FGP is a positive grievance or dispute resolution system, but over time, the policy has made an important contribution to the resolution—largely using informal means—of resolving the grievances, disputes, and complaints of MSU faculty and academic staff.

APPENDIX 1: IMPORTANT AMENDMENTS OF FGP SINCE 1990

1. Of the nine amendments of the FGP in the 1990s, three are significant:
   A. 1995 (and 2002): The creation of a procedure to address jurisdictional matters, including access to the FGP by granting the FGO explicit authority to make policy access decisions based on several tests with FGO decisions subject to appeal to a three-person standing panel. Its intent was to provide a quick solution to jurisdictional disagreements.

   B. 1995: The formal inclusion in FGP language to protect the FGO, hearing panel members, and the parties' counsels in their several roles under the university indemnification policy. This replaces a previous board of trustees' resolution that was not included in the original FGP language.

   C. 1999: The actions (in three separate amendments) to exclude the president, the university general counsel and staff, and the FGO from the jurisdiction of the FGP as grievance respondents. For the Office of the President, this resolved a historic argument about this matter. Also, the changes applied to decisions and actions of the chairpersons of the university Committee on Human Subjects and other internal review boards and to the actions and decisions taken under the Procedures on Allegations of Misconduct in Research and Creative Activities. The various roles and responsibilities of all these individuals and bodies justified exclusion as respondents under the FGP for various reasons. In the case of the various review board decisions, subsequent disciplinary actions taken by university administrators can be grieved under the FGP.

2. Of the twelve amendments of the FGP made since 2000, four are significant:

   A. 2009: The elimination of department/school and college hearing panels providing for initial hearing review only at the provost level. This formalized a traditional FGO practice of encouraging provost-level decisions.

   B. 2009: A new selection procedure for membership of initial hearing and appeals panels was introduced so that all faculty and academic staff with continuing appointment system status or those who had held fixed-term appointments for at least three consecutive years (with certain exceptions) serve with an understanding that no individual can serve on a hearing or appeal panel more than once in seven years. This is a return to the original jury-like selection procedure with a more efficient process.

   C. 2016: Appeal panel provisions were changed to eliminate personal hearing panel presentations by appellants and respondents. This change was to provide more time for appeal panels to consider appeals materials without the need to schedule relatively quick hearings. It also was intended to discourage the parties from retrying their cases in person, such being precluded under FGP provisions. This was not a separate FGP amendment but was part of a general revision of university discipline, dismissal, and faculty grievance policies.

   D. 2018: The FGP policy's name and that of its office was changed to Faculty Grievance and Dispute Resolution Policy and Office. Among other reasons for this change was a desire to advertise a mission change in the FGP with an intent to provide a wider range of services beyond traditional grievance resolution, including conciliation, mediation, and conflict resolution.

Potentially, the inclusion of dispute resolution in the policy's title might encourage more faculty and academic staff to use the FGP's services. Because both items C and D are relatively new, the impact of those amendments is as yet unclear.

APPENDIX 2: SOME SUGGESTIONS FOR FUTURE REVISIONS OF THE MSU FGP

1. In a technical sense, we believe the FGP needs no changes in its rules and procedures. It has been tweaked on many occasions over the years and, on the whole, the policy works well. Issues can be addressed by making ad hoc changes over time. One area of concern is the absence of a determination or at least an estimate of the time taken in deciding or resolving grievances formally or informally. This is one of Estey's evaluation criteria, and on this matter it is suggested to keep a record on each grievance with a timeline starting with the filing of a grievance or an indication of a desire for an informal resolution with an end date of when the grievance is decided either in a formal hearing or via informal resolution. While not without flaws, this information can help determine the FGP's level of efficiency or inefficiency.

2. Data and Survey Issues

    A. An assessment survey that focuses largely beyond just a review of the FGO to an assessment of the FGO and includes faculty/academic staff, administrators, and FGO office staff was developed recently. It provides a comprehensive sorting of responses by those involved with the FGP in a number of ways. Although response rates are relatively low, over time, the responses give some sense of the FGP's implementation, as well as the performance of the FGO and the FGO office staff. These surveys should be continued, and their outcomes should continue to receive the attention of both the UCFA and the Office of the Provost.

    B. Separately it is proposed to reach out to administrators to ask opinions of the FGP beyond but including more in-depth commentary from administrators who are grievance respondents.

    C. Data on those who contact the FGP office by gender, race, and employment status is now included in the more recent user surveys and provides important information. Historically, our impression is that women, faculty and staff of color, and academic staff apart from faculty are underrepresented in use of the FGP. The largest clientele comes from tenure-system faculty. Between 1974 and 2018, the total number of faculty and academic staff increased from 3812 to 5708, which is the pool covered by the FGP. Over this same time period, the number of temporary faculty increased from 576 to 1333, and the number of academic staff—both continuing and temporary—increased from 686 to 2380. But the total of tenure-system faculty—at approximately 2000—has remained largely flat over this time period. These trends have implications for how and to what extent the FGP is responding to potential users.

    D. Ongoing grievance outcome data as reported in this paper is either spotty or non-existent since 2015, especially during the administration of FGO Donohue. While, as noted, this information has deficiencies, it provides a reasonable sense of outcome information. The absence of this information should be addressed by the current FGO and his/her successors. In addition, efforts should be made to articulate and keep information on the definitions used to record grievance outcomes so as to provide a more common understanding of those definitions.

3. The FGO Position—Centerpiece of the FGP

    A. Improve the selection process. FGP members are currently drawn from a pool of tenured faculty appointed at MSU at or above the rank of associate professor. An issue is concern is small application pools and high turnover among FGOs (two FGOs served between 1989 and 2011, but there was high turnover before 1983 and since 2014).

B. It is worthwhile to consider the pros and cons on long-term FGOs and whether long-term assignments are useful in exploring how an environment can be developed to enhance such an outcome.

C. Consider widening the selection pool to include academic staff, MSU retirees (both faculty and academic staff), and outside professionals. The latter could be appointed as executive managers or on five yearly renewable contracts. Such appointees would require relevant background and experience, be willing to gain knowledge of the university setting, and be able to win the trust of the university community and FGP clientele. Neither author is enthusiastic about this option, but it should be reviewed and considered if the internal applicant pools remain small.

D. If the FGO position remains half time, as has been the case since 2012, the Office of the Provost must commit attention to secure a reasonable assignment for the FGO to satisfy performance expectations in roles both as an FGO and as a faculty member.

E. Determine whether the position of the FGO is really a half-time assignment that allows adequate time to pursue other faculty roles. Recently, the Office of the Provost proposed to establish an associate FGO position, to which a woman would be appointed. This is to provide a more gender-friendly environment and to potentially reach out to a larger pool, but as yet is only marginally served component of temporary faculty and continuing and temporary academic staff. The specific role, functions, and relationship between the FGO position and the associate FGO assignment have yet to be worked out.

F. Hold meetings and produce webcasts to share commentary on the various grievances sources and how unit administrators can help resolve or forestall grievances—especially on critical reappointment, tenure, and promotion or other reappointment appointment/appointment issues—to reduce their inappropriate occurrence (for example, former FGO Michael Rubner's writings could be edited, shortened, and scripted with opportunities for role playing; those writings emphasize the need for setting clear expectations and regular review follow-up in probationary appointment periods with steps on how to achieve these objectives).

G. Achieve a greater presence of women and individuals of color in the candidate pool for the FGO position. Since 1972, of a total of 15 FGOs, only 3 women have served as FGO and 2 have been faculty of color (see above regarding the planned creation of an associate FGO position).

H. Develop a training and mentoring program for FGOs via personal support, development of problem-solving expertise, and encouragement to undertake external training related to the FGO's role. In the selection and training of FGOs, inculcate the importance of trust and impartiality in the FGOs role in addition to basic knowledge of the FGP rules and procedures and other administrative responsibilities. In addition, the strength of the FGO's commitment to mediation is very important in light of the significant number of grievances that are addressed via informal resolution. Mediation expertise also is important, as noted previously, because of the existence of a hearing component within the FGP that encourages informal grievance resolution assisted through mediation efforts. In addition, FGO-supported training programs for hearing panel members, counsel, and presiding officers should be continued and enhanced.

I. Require that the FGO regularly attend and engage in UCFA meetings. Such meetings are a source of advice on and general knowledge of university issues and support the case for regular attendance at university council meetings. Involvement with the UCFA also provides a forum where the FGO can bring issues, concerns, and possible changes in the FGP for discussion and resolution. Such interaction also assists UCFA members, who have short appointment terms, to become more knowledgeable about the FGP and their roles and responsibilities regarding the policy. Also, it is important for the FGO to develop a continuing interaction with the associate provost and associate vice president for academic human resources and relevant members of the Office of the General Counsel to provide regular dialog and information sharing.

4. In a number of grievance outcomes, apart from immediate specific outcomes detailed in administrative decisions, there occasionally are directives to unit administrators to make changes in unit operations—for example, to clarify faculty performance expectations and criteria. In grievant survey responses, it was sometimes noted with concern that there is no follow-up to ensure that these more generalized directives are followed. Accordingly, we propose follow-up on such matters by the Office of the Provost in relation to formal FGP grievance hearing decisions and the FGO in relation to informal resolution of grievances.

5. Clarify the relationship between the FGP and several other MSU non-union dispute resolution policies/procedures for faculty and academic staff to ensure an understanding of how these policies and procedures can work together to provide options for dispute resolution. Examples of these alternative dispute resolutions policies/procedures include administrative review (closely linked between the FGP and the Office of the Provost), the role of a UCFA standing committee in addressing cases that may lead to severe disciplinary actions (newly established and to date infrequently used), the complementary role for the FGP beyond the specified contract coverage of the UNFT–MSU collective agreement for temporary faculty, the anti-discrimination judicial board, the complaint procedure for the cooperative extension service, the Facility for Rare Isotype Beams grievance procedure, the role of the university Committee on Human Subjects and other internal review boards, the role of the UCFT in tenure revocation cases, and the procedures covering allegations of misconduct in research and creative activities. The object would be to determine and clarify the relationships between the FGP and these other procedures to ensure knowledge about areas in which the FGP has no role or a limited one and those in which it has an important and collaborative role. Importantly, communication on the status of these relationships should be shared widely across the university, and FGOs should be encouraged to develop good communication and positive relationships with the administrators of these other policies and procedures.

6. The interplay between the FGO's mission, the extent of its jurisdiction, and the policy's capacity to respond to mission requirement should be reviewed. As noted in 2c above, the FGP's jurisdiction has expanded rapidly with varying growth in covered employment subgroups. However, the growth in jurisdictional size has been accompanied by a very small number of FGP users, although the numbers covered are rough estimates. To achieve a better match between the FGP's current user base and its growing jurisdiction, much more extensive contact and advertisement about the FGP's availability is needed, as well better contacts with dispute resolution entities that overlap the same jurisdiction as the FGP.

7. Accompanying this growth in the FGP's jurisdiction, consideration has been given to less emphasis on grievance hearings and a greater focus on problem solving, improved communication, and mediation (occasionally with outside mediators). While this focus has yet to appear, the recent name change in the policy to a Faculty and Dispute Resolution Office appears to encourage such a change. This remains to be seen. We believe it might be useful to go further with a name change—the Faculty and Academic Staff Dispute Resolution Office with the same current access requirements might be considered. A noted, this change in mission is yet to be implemented, but it has implications for the role and responsibilities of the pending associate FGO position. The authors comment that the proposed "new" emphasis on mediation seems curious because mediation has been a major component of the FGO's role historically, in view of the fact that a very large proportion of grievances are settled informally.

8. One aspect of this study is to gather information about non-union grievance policies in existence for faculty and academic staff members employed at the BTAA universities. The preliminary results of this survey reported in the paper shows the existence of wide range of policies and procedures of this kind in

place. The BTAA has from time to time created cross-university study groups on a variety of topics that bring together professionals engaged in the implementation of these topics as a way to share information as a means of supplementing formal training. We suggest that a proposal to establish such an interest group for those administering these policies be referred to the BTAA provost's group for consideration and potential endorsement by the BTAA administration.

APPENDIX 3: DATA

TABLE 1
Jobs of the FGO and People Who Were MSU FGOs

FGO	Service Year	Department/Field
Michael J. Harrison	08/01/1972–07/31/1973	Physics
Bruce L. Miller	08/01/1973–08/31/1976	Philosophy
Charles Patric Larrowe	09/01/1976–08/31/1980	Economics
Frederick D. Williams	09/01/1980–02/28/1982	History
Bruce L. Miller	03/01/1982–12/31/1982 (acting)	Philosophy
Gordon L. Thomas	01/01/1983–07/31/1983 (acting)	Communications
Mary Lee Brady	08/01/1983–12/31/1988	Social Work
Charles Patric Larrowe	01/01/1989–06/30/1989 (acting)	Economics
Michael Rubner	07/01/1989–06/30/2004	James Madison
John Revitte	07/01/2004–08/31/2011	Labor Relations
Linda Jackson	09/01/2011–11/30/2011	Psychology
John Revitte	12/01/2011–06/30/2012 (interim)	Labor Relations
Sheila J. Teahan	07/01/2012–08/14/2014	English
William Donohue	08/15/2014–08/06/2018	Communications
John Bell	08/07/2018–09/17/2018 (responsible administrator)	Education
Francisco A. Villarruel	09/18/2018–12/31/2018 (responsible administrator)	Human Development
Francisco A. Villarruel	01/01/2019–present	Human Development

Evaluation of MSU Faculty Grievance Policy Against Criteria

TABLE 2
MSU Faculty Grievance Procedure Wins/Losses (1972–2015)

Outcome	Type	Number of Cases	Percentage
Grievant Won	Informal	216	19.20%
	Formal	43	3.80%
Grievant Lost	Informal	71	6.30%
	Formal	132	11.70%
Grievant Satisfied	Informal	324	28.80%
Grievant Dropped	Informal	311	27.70%
Others	Informal	27	2.40%
Total		1124	100%

Source: Annual Reports of the Faculty Grievance Official to the Academic Council, Michigan State University.

FGO Annual Report Summary

TABLE 3
Grievance Cases Settled (with Notation of Which Were Settled Informally or Via Formal Grievance Hearings) and Those Cases Still Pending at the End of Each Academic Year

Years	Cases Settled Informally	Formal Hearing	Total Cases Settled	Pending Cases	Total Cases
1972–2018*	1112	186	1329	793	2122
%	52.4%	8.8%	62.6%	37.4%	100%
Average (annually)	25	4	30	18	47

The 2016–2017 data is excluded because the data from fall 2016 is missing.

TABLE 4
Formal Hearing Settlements by Subject

Years	Salary	Evaluation/ Salary	Reappointment	Promotion and Tenure	Promotion	Assignment/ Reassignment	Total Cases
1983–2018*	25	19	13	27	7	8	137
%	18.2%	13.9%	9.5%	19.7%	5.1%	5.8%	100%

The 2016–2017 data is excluded because the data from fall 2026 is missing; the 2014–2015 data and 2015–2016 data are also excluded because the FGO (William Donohue) did not sort the settlements by subject. The 2014 fall semester report does have settlements by subject, but the 2015 spring and summer semesters do not.

TABLE 5
Informal Settlements by Subject

Years	Salary	Evaluation/ Salary	Reappointment	Promotion and Tenure	Promotion	Assignment/ Reassignment	Total Cases
1983–2018*	190	140	57	57	55	104	950
%	20%	14.7%	6%	6%	5.8%	10.9%	100%

The 2016–2017 data is excluded because the data from fall 2026 is missing; the 2014–2015 data and 2015–2016 data are also excluded because the FGO (William Donohue) did not sort the settlements by subject. The 2014 fall semester report does have settlements by subject, but the 2015 spring and summer semesters do not.

Total cases settled by subject and in some years an evaluation by the FGO of whether—from the grievant's perspective—the grievant won, lost, was satisfied with a "compromise" settlement, dropped the case, and so forth.

TABLE 6
Inquiries by Subject Matter

Years	Salary	Evaluation/ Salary	Reappointment	Promotion and Tenure	Assignment/ Reassignment	Grievance and Procedure	Discipline	Bylaws/ Harassment	Total Cases
1984–2018*	182	713	198	375	500	1149	414	432	6323
%	2.9%	11.3%	3.1%	5.9%	7.9%	18.2%	6.5%	6.8%	100%

The 2016–2017 data is excluded because the data from fall 2026 is missing

TABLE 7
Conferences and Meetings Held by Subject

Years	Salary	Evaluation/ Salary	Reappointment	Promotion and Tenure	Assignment/ Reassignment	Grievance and Procedure	Discipline	Bylaws/ Harassment	Total Cases
2004–2018*	84	250	162	281	167	370	179	103	2468
%	3.4%	10.1%	6.6%	11.4%	6.8%	15%	7.3%	4.2%	100%

ENDNOTES

[1]Clifton R. Wharton Jr., Privilege and Prejudice: The Life of a Black Pioneer, Michigan State University Press, 2015, chapter seven.

[2]Statement by the American Association of University Professors—Michigan State Chapter, November 7,1966; Discussion of Faculty Reappointments, April 16, 1971 Michigan State University board of trustees meeting transcript; Herman king, Procedures for Considering Reappointments of Non-tenured Faculty; memo the MSU Council of Deans, October 22, 1971.

[3]The Ad Hoc Committee to Study Rights, Responsibilities and Grievance procedures, The Interim Faculty Grievance Procedure, September 23, 1971.

[4]The Interim Faculty Grievance Procedure, Approved by the MSU Board of Trustees, May 19, 1972.

[5]Tape recording of the MSU Academic Senate meeting, November 5, 1975.

[6]Memo dated February 27, 1979 to the Faculty Council dated entitled Faculty Grievance Procedure by Provost C.L. Winder. An earlier draft entitled Faculty Grievance Procedure which compared the language of the original FGP, the UCFA proposal and the Faculty Council proposal dated February 6, 1979 was sent to Faculty Council by Provost Winder.

[7]Minutes of the MSU Faculty Council meeting dated November 13, 1979 referencing a memo from Provost C.L. Winder entitled Revising the Interim Faculty Grievance Procedure, dated November 5,1979. Memo from President Cecil Mackey entitled Outline for a Faculty Grievance Procedure dated February 19,1980 attached to a cover letter from Provost C.L. Winder to the Chair of the Steering Committee of Academic Governance dated February 27, 1980.

[8]Douglas Noverr, Michigan State University: The Rise of a Research University and the New Millennium, 1970–2005, Michigan State University Press, 2015, pp. 66–67 and 110–16.

[10]Marten Estey "The Experience at AAU Campuses," *Academe* May–June 1986. pp. 6–15. This essay is based on a study supported by a grant from the Exxon Education Foundation and co-sponsored by the American Association of University Professors (AAUP) and the Association of American Universities (AAU). This study of AAU members makes its focus relevant to MSU as an AAU member. It examines the characteristics of and experience with non-union faculty and academic staff grievance procedures at about 50 campuses as AAU members in the mid 1980s which vary in size and scope but which in general make up the lion's share of American graduate and research universities and are composed of a mix of public and private institutions. Information for the study was gathered by survey and a number of telephone and personal interviews with relevant individuals on the various campuses.

SOURCES CONSULTED

In addition to the Estey article referenced in this paper, additional literature on the topic listed below was reviewed.

Begin, James P. 1978. “Grievance Mechanisms and Faculty Collegiality: The Rutgers Case,” ILR Review, Vol. 31, No. 3, pp. 295–309.

Benewicz, Maurice C. 1973. “Grievance and Arbitration Procedures” in Terrence Tice, ed. Faculty Bargaining in the Seventies, Ann Arbor, MI, Institute of Continuing Legal Education.

Conlin, Donald E. et al. 2007. “Third-Party Interventions Across Cultures: No ‘One Best Choice,’” Research in Personnel and Human Resources Management, Vol. 26, pp. 315–356.

Eaton, Adrienne, and Jeffrey H. Keefe (eds). 1999. Employment and Dispute Resolution and Worker Rights in the Changing Workplace, Industrial Relations Research Association Series.

Hannah, John A. 1980. A Memoir, East Lansing, MSU Press

Jacoby, Sanford M. 1999. Modern Manors: Welfare Capitalism Since the New Deal, Princeton, Princeton University Press.

Kuhn, Madison. 1955. Michigan State: The First Hundred Years, East Lansing, MSU Press

Peterson, Richard. 1992. “The Union and Nonunion Grievance System, Research Frontiers in Industrial Relations and Human Resources, Industrial Relations Research Association Series, pp. 131–162.

Repas, Bob. 1977. “Grievance Procedures Without Arbitration,” ILR Review, Vol. 20, No. 3, pp. 381–390.

Rowe, Mary. 1997. “Dispute Resolution in the Non-Union Environment: An Evolution Toward Integrated Systems for Conflict Management?” In Sandra Gleason (ed), Workplace Dispute Resolution, East Lansing, MSU Press, pp. 79–106.

ADDITIONAL NOTES

The source of projecting the numbers of FGP users is drawn from the annual survey reviews and from the five yearly reviews of the FGP and the FGO. However, many annual review forms have been misplaced. Under the online survey system introduced after 2000, in a number of cases of responders to the surveys is provided—not the total number sent to users in a particular year—are included the response rate for these surveys runs about 50%. In the past, although not regularly, some individual FGOs kept records of participant numbers. All this information was considered in making a projection of the average annual number of participants as 150. But bear in mind these are rough projections.

The FGO annual reports can be reviewed at the MSU Archives and Historical Collections.

Revitte reviewed and compiled the data from annual reports used in the tables presented in Appendix 3.

The authors interviewed presidents Lou Anna K. Simon (in person) and M. Peter McPherson (via telephone).

V. Exploring the Path to Becoming a Successful Employment Law Mediator and Arbitrator

Labor Arbitration and Employment Arbitration: Are They Twins or Just Cousins? An Overview for Aspiring Employment Arbitrators

JACQUELIN F. DRUCKER, ESQ.
Drucker Arbitration

Introduction

Many neutrals and aspiring neutrals who have worked primarily in the labor-management arena are considering the possibility of expanding their practices into employment arbitration, which involves alleged violations of federal, state, and local employment laws or breaches of individual employer–employee contracts. To provide some guidance for those navigating such an expansion or transition, this discussion highlights a few of the key differences in the two types of arbitration practices.

Differing Characteristics of the Arbitrators

Most labor-arbitration rosters, such as those of the American Arbitration Association (AAA) and the Federal Mediation and Conciliation Service (FMCS) are populated by individuals who have had significant experience in labor–management relations or labor law and who do not currently function as advocates for unions or employers. By contrast, most employment-arbitration panels, such as that of the AAA, consist primarily of those with experience in employment litigation. As a result, a significant majority of employment arbitrators are attorneys and, unlike the overwhelming majority of established labor arbitrators, some maintain active law practices representing employers or employees. Thus, for those who hope to become neutrals but currently are working as advocates, employment arbitration (and, even more so, employment mediation) can present opportunities to serve even before stepping away from the representation of clients.

Differences in the Arbitrators' Disclosure Obligations

In both forms of arbitration, neutrals have an ongoing obligation to disclose connections they, their families, or their staff have or had to the parties, attorneys, law firms, or entities. In this regard, labor arbitrators are bound by the Code of Professional Responsibility for Arbitrators of Labor–Management Disputes (Labor Code), whereas employment arbitrators in most cases are bound by the Code of Ethics for Arbitrators in Commercial Disputes (Commercial Code). The principles set forth in each code are similar, requiring disclosure of any connection that might taint the perception of the arbitrator's impartiality. In practice, however, the disclosure obligations play out very differently, especially as to the scope of disclosable "relationships." Labor relations communities tend to be professionally interactive and congenial, and labor arbitration often involves advocates, parties, and arbitrators who have worked together for years, through ongoing processes under collective bargaining agreements. As a result, any familiarity or contacts an arbitrator may have with counsel and the parties through prior service as a neutral or involvement with professional organizations are not considered to be relationships that require disclosure (Formal Advisory Opinion No. 22 [1991], National Academy of Arbitrators). By contrast, such connections are specifically and wisely required in employment arbitration, which arises in situations where there seldom are ongoing arbitration-related contacts among the parties, counsel, or arbitrator, and the parties may be strangers to the arbitral forum. Most notably, employment arbitrators are expected to disclose prior service as a neutral with any of the parties, attorneys, or law firms. (Any arbitrator who plans to hear employment cases in California must review and follow the highly specific disclosure requirements under the California Code of Civil Procedure.)

To fulfill these expanded disclosure obligations in employment cases, an arbitrator must establish and maintain a database and other systems that can be searched to identify prior cases or connections that give rise to disclosable connections.

Financial Matters: Who Pays and How Arbitral Services Are Billed

By far the most common practice under collective bargaining agreements is for the employer and the union to share equally the labor arbitrator's fees. The question of who pays for the arbitration fees in employment arbitration, however, has been more complex. At one time, many employment arbitration plans, often those to which agreement is required as a condition of employment, provided that the fees of the arbitrator were to be shared equally by the parties, requiring an employee to pay for a forum that would have been free if the claim had been pursued in court. These provisions, however, have been found in many jurisdictions to be problematic and unenforceable (see, e.g., discussion and review of cases in *Musnick v. King Motor Co.*, 325 F.3d 1255, 1259 [11th Cir. 2003]). This led employers to move away from fee-sharing plans and caused organizations such as AAA to require, through rules and administration, that employers assume responsibility for paying the fees and expenses of the forum, with the exception of a filing fee.

Employment arbitration thus creates a different financial dynamic for the arbitrator. Advance deposits to cover arbitrator fees are common. In addition, unlike the typical per diem rates and billing used in labor arbitration, employment arbitrators typically bill at an hourly rate and are expected to produce itemized invoices in hourly increments.

Pre-Hearing Activity

A labor arbitrator often enters the hearing room having only the most general sense of the issue in dispute. By contrast, in employment arbitration, the arbitrator begins involvement early by reviewing the initiating pleadings, conducting pre-hearing management conferences, and, at times, entertaining various pre-hearing motions.

In addition, employment arbitration often involves elements of formal discovery. In labor arbitration, the grievance process provides for exchange of information and arguments, and, thus, the discovery mechanisms of litigation, such as depositions, interrogatories, and formal requests for production, are generally not used in labor arbitration. In employment arbitration, however, conventional discovery methods often are used, albeit in a way that is more limited than in litigation. This means that the arbitrator in an employment matter must manage the pre-hearing processes and be prepared to resolve disputes regarding the degree to which discovery will be used, in keeping with the expedited and streamlined nature of arbitration, and whether the parties have complied with whatever discovery has been authorized.

Types of Remedies

Remedies in labor arbitration are contractual in nature and usually limited to make-whole relief. Punitive damages, awards of interest, and payment of attorney fees are not available, absent special contract language or agreement. By contrast, the range of remedial relief in employment arbitration is governed by the applicable statutes and contracts. Thus, in appropriate cases, punitive damages may be warranted. It also is common for attorney fees to be awarded, in keeping with statutory requirements. Employment arbitrators must be prepared to award such fees when the law or contract requires it and to provide the mechanism for resolution of disputes regarding entitlement or calculation.

Opportunities

Labor arbitrators seeking to expand their practices to include employment arbitration can enhance the process by highlighting or, where necessary, updating their working knowledge of employment law and litigation techniques. Involvement in the development of substantive programming and materials within professional organizations often is useful in this regard. It also can be productive to begin mediating employment disputes, the opportunity for which may be available through the courts, albeit often *pro bono*. In addition, some limited options exist for employment arbitrator training, such as occasional seminars offered by the Scheinman Institute on Conflict Resolution at Cornell University's ILR School. For more information and ideas, email the author at jdrucker@druckerarbitration.com.

VI. New Developments in Minimum Wage Research

An Earnings Standard for New York City's App-Dispatched FHV Drivers: Economic Analysis and Policy Assessment

JAMES A. PARROTT
The New School

Abstract

New York City's Taxi and Limousine Commission implemented a $17.22 after-expense per hour minimum driver pay standard for app-dispatched FHV drivers in February 2019. The pay standard was the first in the nation for app drivers. Prior to the standard, 72% of all trips paid drivers less than $17.22 per hour, and 96% of all drivers had one or more trips per week paying less than the minimum. The pay standard formula incentivizes companies to make more efficient use of drivers' time, reducing the time spent cruising while waiting for the next dispatch.

Introduction

Four app-based companies—Uber, Lyft, Via, and Gett/Juno—have expanded rapidly in New York City since 2012. The app-based industry includes about 85,000 vehicles (as of the end of 2018), dwarfing the city's 12,000 medallion (Yellow) taxi fleet.[1] App-based drivers now complete over 20 million trips in the city each month, two-and-a-half times the number of medallion trips. [2]

App car service growth has been rapid—the number of trips rose by 122% in 2016, 72% in 2017, and 47% in 2018. The average number of daily trips by app services was about 100,000 in June 2015, around the time New York City leaders first raised the idea of a cap on app service growth. That suggestion was resoundingly thwarted by an aggressive Uber public relations campaign. By March of this year, the number of average daily app service trips had soared to nearly 800,000.

As Figure 1 (next page) illustrates, the growth in app services has taken considerable market share from medallion taxis, but they have also tremendously added to the combined number of daily taxi and app service trips. The effects on congestion in the Manhattan central business district are inescapable, but the app services have also spread to almost every neighborhood in the city. The combined number of taxi plus app trips doubled between March 2015 and March of 2019 when the average number of trips topped one million.[3]

The industry provides more jobs (on a full-time equivalent basis) than many prominent industries, including commercial banking, hotels, and publishing. Uber alone would be the largest for-profit private employer in New York City—if Uber drivers were classified as employees rather than independent contractors.

This rapid growth has generated substantial benefits—including increased convenience for riders and the extension of transportation services to neighborhoods in the outer boroughs that are not well-served by mass transit. The industry has also generated high returns for its pre-IPO investors and created many new jobs for drivers. But it has also created several problems, including increased competition for the traditional yellow taxi sector and its drivers, downward pressure on FHV driver earnings, and their rapid growth has contributed to worsening congestion in midtown Manhattan. Growth of the app-dispatched FHV services has also diverted passengers from mass transit, reducing revenues at a time when the system needs more investment to modernize antiquated equipment and improve service quality.

FIGURE 1
Average Daily NYC Trips, 2013–2019

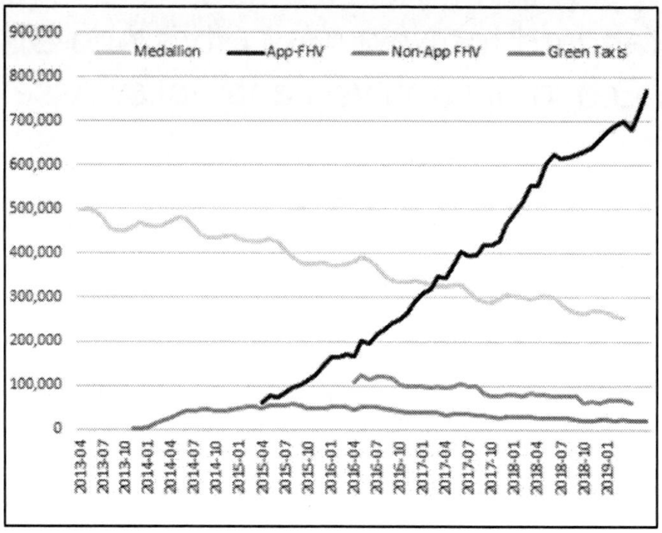

Source: NYC Taxi and Limousine Commission, data through March 2019.

As app service growth soared, congestion worsened, FHV driver pay dropped, and the pressure on the yellow taxi sector and all drivers became so intense that eight drivers committed suicide in the span of roughly a year from late 2017 to late 2018. These developments resulted in significant regulatory interest on the part of the NYC Taxi and Limousine Commission (TLC) in establishing a minimum driver pay standard, and in renewed interest in City Hall in pausing the rapid growth in the number of cars on the streets to allow time for a comprehensive analysis of the impact on congestion and related issues.

The app-dispatch driver pay standard implemented this past February was the first in the nation and affects over 85,000 independent contractor drivers. The app-dispatch industry is by far the largest segment of the online platform "gig economy." The New York City TLC pay standard will be closely watched for its implications for the app industry in other major cities and generally for lessons regarding establishing worker protections in the "gig economy."

The App-Dispatch Business Model

In the app-based FHV industry, the drivers are independent contractors, not employees.[4] The companies set the fares and the number of new drivers using their apps. The drivers set their own schedules and total number of work hours and are paid a share of the revenue generated by their passenger trips. However, driver payment has not always been a fixed proportion of the passenger fare, and depended on various company policies, such as promotions for drivers and riders, treatment of shared rides, and route-based pricing. Since the February implementation of the driver pay standard, the standard also affects the amount of driver pay.

Drivers, not the companies, provide the bulk of capital investment. The drivers supply the vehicles and pay for all their driving-related costs (vehicle licensing, insurance, maintenance and repairs, and fuel) that we conservatively estimated at $22,000 annually for the weighted average of vehicles in use in 2018 (Parrott, Reich, Rochford and Yang 2019). For a car driven 35,000 miles a year, that works out to 63.1 cents per mile.[5]

Sixty percent of app-based drivers are full-time workers who undertook risky capital investments in the vehicles they acquired to drive in this industry. These drivers, 90% of whom are immigrants and most lacking a college education, have difficulty obtaining better-paying jobs elsewhere in the New York economy. Nearly 90% of non-four-year college degree immigrant males ages 25–44 working in New York City hold blue collar, lower-paying white collar, or service jobs in 2016 with median annual earnings between $25,000 and $28,000. Once they have committed to acquiring a vehicle, they face high exit costs if they discover that it is difficult to cover all their expenses and net a reasonable amount of after-expense earnings. TLC data indicate that over one-quarter of new app drivers leave within their first year, rising to 35% by the end of two years (Parrott and Reich 2018).

Nearly one-fifth of New York City FHV and taxi drivers receive Supplemental Nutritional Assistance Program aid (food stamps), compared to about 10% of the overall local workforce. One in six had no health insurance coverage, and 40% were covered by Medicaid (Parrott and Reich 2018).

Prior to the pay standard, the variability in hours of existing drivers and the steady recruitment of new drivers allowed the companies to play the dominant role in determining driver pay. The companies compete with each other primarily by minimizing passengers' wait times and by keeping fares low, particularly for price-sensitive customers. They compete also by expanding their geographic coverage with more vehicles to service a larger customer base, especially in areas of New York City that are under-served by mass transit and yellow taxis. To achieve quick response times, the companies have typically required many idle drivers to be available at any given moment and at many locations.

This app-dispatch business model created a gap between the drivers' desires to maximize their earnings—by maximizing trips per working hour—and the companies' desire to minimize response times. In other words, before the pay standard, the app business model relied on keeping driver utilization low, which then kept drivers' hourly pay low as well.

The companies receive a commission on every trip that varies from 20-30% (sometimes more). Given relatively moderate corporate overhead costs (maintaining the app, driver recruitment costs, credit card and bank fees, and advertising, legal and lobbying costs) relative to the revenues generated, we estimated that for 2017 and 2018 the net profits received by Uber and Lyft (which together accounted for nearly 90% of the market) were considerable (Parrott and Reich 2018).

Analysis of Driver Net Earnings before the Pay Standard

Utilizing extensive TLC administrative data on all app drivers for four study weeks between September 2016 and October 2017, we determined that, on a per-trip basis, drivers received less than $17.22 an hour for 72% of all trips after covering expenses that we estimated at $0.63 per mile. Ninety-six percent of all drivers had at least one trip that paid less than the $17.22 minimum during the mid-October 2017 study week. As indicated in Table 1, we estimated that the median after-expense hourly earnings for app drivers in mid-October 2017 was $13.70.

TABLE 1

Summary, Per-Trip Earnings Analysis Results* (expense factor: 63.1 cents per mile)

	Uber	Lyft	Juno	Via	Combined
All trips					
# of trips	2,162,597	628,460	245,131	18,268	3,054,456
Mean gross hourly pay	$23.10	$23.54	$25.27	$31.64	$23.42
Median gross hourly pay	$21.85	$22.58	$23.34	$31.12	$22.18
Mean after-expense hourly pay	$14.67	$14.72	$16.88	$22.94	$14.90
Median after-expense hourly pay	$13.56	$13.78	$14.24	$22.48	$13.70
Trips below $17.22 minimum					
# of trips below minimum	1,587,482	463,866	171,169	1,672	2,224,189
% of trips below minimum	73.4%	73.8%	69.8%	9.2%	72.8%
Mean gross hourly pay	$19.85	$20.48	$21.84	$19.37	$20.14
Mean after-expense hourly pay	$11.08	$11.59	$13.12	$10.75	$11.34
Utilization rate	58%	58%	50%	70%	

*Based on 4th study week, Oct. 16–22, 2017.

To raise all below-minimum trips up to the minimum standard level would entail a $20.75 increase in weekly aggregate gross pay for all drivers.

For all affected drivers with at least one trip paying below the minimum, we estimated that the pay standard would entail an average 24.7% increase in mean gross hourly pay from $22.35 before the pay standard to $27.86 after the pay standard (Parrott, Reich, Rochford and Yang 2019).

Driver Pay Standard

The driver pay standard formula enacted by the TLC in December 2018 that took effect this past February 1 combines an expense component and a time component. The standard was initially set at $17.22 hourly rate as the independent contractor equivalent of $15.00 (the state minimum wage applicable in New York City as of December 31, 2018 for employers with 11 or more employees), plus 90 cents for a moderate amount of paid time off. Independent contractors must pay the employer share of federal payroll taxes (7.65%). This payroll tax rate applied to $17.22 equals $1.32, which when subtracted from $17.22 yields $15.90 ($15 plus 90 cents for paid time off.) The paid time off factor was derived by multiplying 6% times $15.00.[6] The paid time off rate of 6% is the average cost of paid time off relative to wages for workers in transportation occupations nationwide, according to the Bureau of Labor Statistics (Bureau of Labor Statistics 2018).

The expense component was estimated at $0.631 per mile and is intended to allow the typical driver to cover all the costs of acquiring and operating a vehicle (as well as the cost of licensing and training).[7] The $0.287 per minute factor is intended to compensate drivers for their time at $17.22 an hour ($0.287 is $17.22 divided by 60 minutes). See Figure 2.

Both the expense and time components are divided by utilization because drivers are working even when they do not have a passenger in their car. Driver working time is measured by the time they are logged on to a company's app, and therefore available to carry passengers. Their work time includes the time they are using their vehicle, and incurring expenses for doing so, even when they are cruising while waiting for a dispatch or heading toward a pickup location after having accepted a ride request or returning from a drop-off location.

FIGURE 2
Driver Pay Standard Applied to a Typical Trip

A typical FHV trip might be 7.5 miles in distance and 30 minutes in time.

Here is how the driver minimum pay standard (not the passenger fare) would be calculated under the proposal pay standard (assuming an industry-wide average utilization of 58%):

$$\frac{(.631 * 7.5\ miles)}{.58\ utilization} + \frac{(\$0.287 * 30\ minutes)}{.58\ utilization} = \$23.00$$

Note: this is the minimum pay standard (for a non-shared ride)*, not the passenger fare. The company and the driver can always agree that driver pay for any trip should be higher.*

The driver pay standard ensures that the driver can cover vehicle expenses as well as get paid at least the independent contractor equivalent of $15.00 an hour. Both the per mile and the per minute pay standard components are indexed by the New York metropolitan area Consumer Price Index beginning January 1, 2020.

The pay formula is thus constructed to compensate drivers for work-related time and expense when a passenger is not in the vehicle. It does so by dividing the expense and time components by company-specific utilization rates—as measured for each company by the TLC for some baseline period. However, because of concerns raised by some of the companies, the TLC agreed to utilize the industry-wide 58% utilization factor for the first year and then switching to company-specific utilization factors after that. In the case of the time factor, the utilization rate adjusts for the portion of each hour that a passenger is not in the vehicle. For the expense factor, the utilization factor adjusts for the expenses associated with pickup, cruising, and other non-passenger vehicle uses during the work shift.

This part of the policy in effect incentivizes each company to raise its company-wide utilization rate, that is, by increasing the average number of trips per driver hour. For 2017 and 2018, the TLC determined that the average company-specific utilization rates were those indicated in the last row of Table 1.

Since the utilization rate appears in the denominator, a higher company utilization rate lowers the company's costs for the expense and time components. At the same time, company policies that increase utilization rates will also benefit the drivers. They will be able to provide more rides in any given hour, thereby earning more on an hourly basis, even though their pay for each trip might be lower. Increases in driver utilization rates represent an improvement in industry efficiency. With greater efficiency, the policy better aligns the interests of the drivers with that of the companies and both sides benefit.

Adjustment to the Pay Standard

In analyzing the likely impact of the pay standard, we modeled the extent to which drivers would change their labor supply, how the companies might respond to higher driver pay, and in turn, how consumers were likely to respond should there be a change in fare policy.

With higher pay per trip, drivers on average will increase their labor supply. But this will vary depending on the hours they usually drive and individual considerations. Drivers currently working very long hours, including the 22% of drivers working 50 or more hours each week, are likely to reduce their hours if their effective hourly pay rises. Other full-time drivers and many part-time drivers may respond to the earnings increases per hour by increasing their working hours. Overall, we used an average labor supply elasticity of 0.4.

We also made plausible assumptions about how much companies would increase their utilization of drivers' time. Because of the structure of the pay standard formula, this increase in driver productivity would absorb a large part of the cost of the driver pay increase. The portion of the pay increase not absorbed by increased utilization could be accommodated through a combination of adjustments to passenger fares and company commissions. Considering that services are likely close to a saturation point in core Manhattan, we relied on a consumer elasticity of –1.2 more reflective of demand in the boroughs outside of Manhattan where most household incomes are not as high as in Manhattan. Given likely rider resistance to higher fares, we hypothesized that the companies would reduce commissions to maintain market share. Higher utilization could also have an effect in increasing passenger wait times, but it seemed that wait time increases would be 15–30 seconds at most.

From our modeling that simulated alternative adjustment scenarios, we concluded that the pay increase likely could be absorbed through a combination of utilization increase, commission reduction, modest fare increase, and very slight increase in passenger wait times.

Very Preliminary Results from the First Three Months of the Pay Standard

The pay standard took effect February 1, 2019, and based on data received from the companies through the end of April, the TLC reported that driver pay increased by more than $150 million compared to what it would have been without the pay standard for the first three months. While there was news coverage of increased fares in the immediate wake of the pay standard taking effect, it appears that the largest companies provided extensive fare discounts and that net passenger fares were little changed for this initial period. TLC data indicate that the number of trips continued to increase. It appears that the pay standard is having the desired effect of encouraging the companies to increase utilization and make more efficient use of drivers' time. This was also borne out in testimony from drivers at an April TLC hearing.

In May of this year, both Uber and Lyft announced that they had stopped on-boarding new drivers in New York City. It seems likely that this is partly the result of the pay standard, and partly the result of the cap on new app-dispatch vehicles that was put in place in mid-August of 2018. Increased utilization and limiting the number of new drivers both indicate that the companies are more closely managing their use of drivers—a positive development that should augur well for driver pay.

Since a considerable wave of new vehicle registrations occurred in the two weeks before the effective date of the cap, the number of high-volume FHVs in use and the number of drivers increased about 11% in the six months following the cap compared to the prior six months (that's about 1,400 additional cars each month.)

Given the importance of the New York City app-dispatch market to the industry's broader fortunes, the impact of the minimum driver pay standard will be closely watched. And since the app-dispatch car services are such a large part of the online platform economy, New York City's regulation of independent contractor driver pay should hold lessons for "gig economy" worker practices more generally.

What's Next on the Regulatory Front?

The City Council legislation establishing the 12-month cap on the number of new FHVs called for a study by the TLC and the City Department of Transportation to analyze the effects of FHV growth on congestion, safety, and medallion taxi and FHV driver income and well-being. The study is intended to inform the development of a TLC FHV growth management policy that would be implemented following the expiration of the one-year pause.

In related developments, the City Council has created a task force to study taxicab medallion values, the TLC has reduced the amount that medallion drivers can be charged for credit card fees, an office of inclusion was established within the TLC to address discrimination by drivers, and financial counseling for FHV and medallion drivers has been expanded.

It is not clear how the taxi and FHV industries will be affected by a congestion pricing plan authorized by New York State in March. Fees that could range as high as $12–$14 per car and $25 per truck for vehicles entering a congestion zone in Manhattan south of 60th beginning in 2021. The congestion fees are intended to generate $1 billion a year in funding for New York's mass transit system (Paybarah 2019). Beginning February 2, 2019, a $2.50 MTA surcharge was applied to each taxi trip and $2.75 for each FHV trip (75 cents for a shared trip) traveling partly or wholly within the core Manhattan area. The MTA surcharge was enacted last year. Drivers have been lobbying to eliminate the surcharge.

Endnotes

[1] The number of medallion taxis in service peaked at around 13,500 in 2014.

[2] This presentation draws from two recent reports I co-authored that were prepared for the NYC Taxi and Limousine Commission. James A. Parrott and Michael Reich, An Earnings Standard for New York City's App-Based Drivers: Economic Analysis and Policy Assessment, July 2, 2018. http://www.centernyc.org/an-earnings-standard; and James A. Parrott, Michael Reich, Jason Rochford, and Xingxing Yang, The New York City App-Based Driver Pay Standard: Revised Estimates for the New Pay Requirement, January 2019. https://bit.ly/2wMxN6t

[3] Data from the NYC Taxi and Limousine Commission website. Green taxis are street-hail taxis restricted to areas within New York City outside of Manhattan below East 96th Street and West 110th Street. For Hire Vehicles (FHV) cannot accept street hails and are dispatched through an app, phone, radio or computer. Non-app FHVs include liveries, black cars, and luxury limousines.

[4] Status as independent contractors has been challenged by drivers represented by the New York Taxi Workers Alliance. In its SEC Form S-1 Registration Statement filed prior to its recent Initial Public Offering, Uber noted that if it were required to classify its drivers as employees rather than independent contractors it would "incur significant additional expenses for compensating Drivers, potentially including expenses associated with the application of wage and hour laws (including minimum wage, overtime, and meal and rest period requirements), employee benefits, social security contributions, taxes, and penalties. Further, any such reclassification would require us to fundamentally change our business model, and consequently have an adverse effect on our business and financial condition." Uber Technologies, Inc. SEC *Form S-1 Registration Statement*, April 11, 2019, p. 28.

[5] That is fairly close to the 58 cents per mile 2019 Internal Revenue Service business mileage allowance. New York City FHVs are required to have commercial auto insurance and there are about $1,500 in driver and vehicle licensing, registration and related costs annually.

[6] The amount of the pay standard will be adjusted annually based on the change in the New York metropolitan area Consumer Price Index, with the first adjustment scheduled to occur on January 1, 2020. The New York State minimum wage statute does not provide for indexation.

[7] An expense factor of $0.803 per mile was estimated for wheelchair-accessible vehicles since these vehicles must be modified to provide for a wheelchair lift.

References

Bureau of Labor Statistics. 2018. Employer Costs for Employee Compensation, December 2017.

Parrott, James A., and Michael Reich. 2018. An Earnings Standard for New York City's App-Based Drivers: Economic Analysis and Policy Assessment. New York: Center for New York City Affairs; and Berkeley, CA: Center for Wage and Employment Dynamics.

Parrott, James A., Michael Reich, Jason Rochford, and Xingxing Yang. 2019. The New York City App-Based Driver Pay Standard: Revised Estimates for the New Pay Requirement. New York: Center for New York City Affairs; and Berkeley, CA: Center for Wage and Employment Dynamics.

Paybarah, Azi. 2019. "Congestion Pricing: Who Pays and Who Doesn't." *The New York Times*, April 18.

Uber Technologies, Inc. 2019. SEC Form S-1 Registration Statement.

VII. 2018 LERA Lifetime Achievement Awards

MARLENE HEYSER

Remarks by David Lewin

I've been asked to accept the LERA Lifetime Achievement Award on behalf of Marlene Heyser, who is unable to travel to Cleveland for the 2019 LERA annual meeting. I gladly agreed to do so. Here is a brief "bio sketch" of Marlene. I'm indebted to Bonnie Castrey, former president of LERA, for contributing to it.

Marlene joined OCLERA (then OCIRRA) in the late 1970s as a young practitioner at the Orange County (California) Transit Authority (OCTA).

She continued her education in labor relations over the years and rose from an administrative assistant to the director of labor relations at the OCTA. In that role, she negotiated with all of the unions at OCTA and was one of the first women to have this position of authority. Concurrently, she became active in the national IRRA, impacting the content of the national meetings by strategizing and advocating to have more sessions for practitioners.

For the past 25 years, Marlene has co-chaired the OCLERA Labor Law Conference and thereby significantly impacted all of OCLERA's programming.

Nationally, she served as president of LERA and thereafter continued her passion for building the organization by working on the restructuring of LERA and moving the annual meeting to late spring/early summer, while maintaining the strong winter meeting in conjunction with ASSA.

As an active member of the National Chapter Advisory Committee, Marlene brought the national LERA message to many local chapters, including through several speaking engagements at many chapters, but especially at TERRA conferences. Recently, she served on the LERA By-Laws Committee to help provide increased chapter involvement in LERA. She has also served on the LERA Strategic Thinking Committee and the LERA Media Awards Committee.

In addition to all of her work with LERA, Marlene has been very active in the Labor and Employment Section of the American Bar Association.

After retiring from OCTA, she founded her own business through which she has helped many companies, non-profit organizations, and public agencies improve their workplace practices.

As one of the pioneering women practitioners in our field, Marlene has also mentored many women and opened doors for them in many traditionally male-dominated industries, especially transportation.

When told that she would be receiving the prestigious LERA Lifetime Achievement Award, Marlene was thrilled and read the letter several times, smiling and saying, "Thank you." But of course, she does not have to thank us; rather, it is we who have to thank her. We do so (in part) by presenting her with this singular award, which is the highest honor that LERA can bestow on one of its members.

Because Marlene is unable to travel to Cleveland for the 2019 LERA annual meeting, Bonnie Castrey and I will deliver the award in person to Marlene in Southern California later this year.

HOYT WHEELER

Good afternoon and thank you. I must start my talk today with an apology. I'm not here with you, but rather I'm on a boat on the Danube River in Europe because by the time I found out about this, I'd locked myself into a cruise in Europe, and when I tried to change it, I found that I really couldn't do that without some extraordinary financial costs. So here I am, onscreen. My son, Jeff, is here to accept this on my behalf.

I think what I should do at this point is to give credit to people and institutions that helped me achieve whatever I have achieved in this career of mine. My interest in labor issues began when I was in college at Marshall College, when the National Intercollegiate debate topic was right-to-work-laws. And I, along with my partner—who was way later my wife, Janice—competed in a tournament debating right-to-work law. I was fascinated by the issue. When I got to law school at the University of Virginia, I took a labor law course and arbitration course from Charles O. Gregory. He was one of the old War Labor Board arbitrators, and really a master of the art of arbitration.

I went from law school back home to West Virginia—to Charleston, West Virginia. I was an associate and then a partner at a law firm there, and a large part of my practice was that of a management-side labor lawyer. Labor law as practiced in West Virginia, at that time in the '60s, was very different than it is practiced generally now around the country. West Virginia was 45% unionized—a union density rate of 45%.

In that milieu, you didn't engage in scorched earth. We battled one another in rooms in arbitration cases, across bargaining tables, and sometimes on picket lines, but there was always a good deal of mutual respect. And management lawyers, in order to be effective, had to be able to get along with their union opponents. In fact, it turned out that when I began arbitration, many years later in Wyoming, the Rocky Mountain west, the way I got started in arbitration was on the basis of references from the union people that I had worked with, and against, in West Virginia, and that's really how I got started doing the arbitration work that I've been doing now for about 40 years.

The work as a management lawyer was mentored by a man named Vincent Chaney, who was perhaps the best lawyer that I've ever known. He taught me well. But I also learned that I was very interested in the subject, in studying it and teaching about it. So I went back to school. I went to the University of Wisconsin to get a Ph.D. in industrial relations. As Tom Kochan said last year when he received this award, Wisconsin is really the center of my being in the field of industrial relations. From the faculty, including such diverse people as Jack Barbash for Theory and the Understanding of Trade Unions, to Don Schwab, who taught Clarity of Thought and Rigor in Research. And other professors—Jim Stern, Jim Jones, Rick [unintelligible], and Mike Aiken—were really very, very important to my life.

But I think of even more importance were my fellow students. We learned a great deal from each other, and we worked together after we got our degrees. Tom Kochan and I have written some things together. Roy Adams, Allen Ponak, George Ogle, and then after I left Wisconsin and came to South Carolina, a more recent Ph.D. graduate, Brian Klaas, came to South Carolina, and he and I have done a fair amount of work together as well.

The family part of my history is complicated but important. For many years, my first wife, Janice, was the wind beneath my wings as we went through law school, starting at a law practice—being paid virtually nothing—graduate school, starting out as an academic, even going to the high plains of Wyoming (which she did not like). Her loss was a great loss to me. My present wife, Liz, and I have been married for 23 years. I spoke to an old friend at one of the international meetings who had married again late in life, and he said, "Ah, yes, the deuxième printemps, the second springtime," and I like that description, and I thank my dear Liz for being the second springtime in my life.

I can't let this chance to speak to you folks pass without talking to you for a little bit about what I think is a very, very important issue. I believe that the decline of the labor movement in this country is a catastrophe. It is something that should not have been allowed to occur and something that must be reversed. The problem of not having a very strong labor movement is an economic problem for millions of workers, who could have the strength of working together with their brothers and sisters, who have real power to change the conditions of their working lives. Instead, they have what the old organizing song called "the feeble strength of one" and therefore have almost no power with regard to the organization in which they work.

This also has political implications. I am deeply afraid that we will lose our democracy. We have a president who has no understanding of and cares even less about the institutions of our democracy, and says that he falls in love with a Korean dictator. This is not acceptable, it's not tolerable. It must be changed, and the labor movement is one of the ways that it might be changed. It's almost impossible to have a lively progressive movement in this country without a lively labor movement. If you want to know why wages have not gone up, look to the decline of the unions. If you want to know why Hillary Clinton didn't carry several Midwestern states, look at the decline of unions in those states. Wisconsin, in a very short time, went from 30% union density to 9% union density, and there are similar experiences in other states. Generally expect workers to vote about 60% for the people their unions endorse. You can see the importance of this problem with politics, playing out because of the decline of the labor movement.

As Vladimir Ilyich Lenin asked, "What is to be done?" What is to be done, I believe, is to make major revisions in our national labor relations laws. The Taft–Hartley Act, passed in 1947, based upon unions becoming too strong, can hardly be maintained now that the private sector unions have less than 7% of the labor force. This law is obsolete. It needs to be gutted and torn out of national labor relations law. It has become what Harry Truman called when he vetoed it, "a slave labor act"—not that when it began, but has become that over the years. The changes that need to be made are major changes. The attempts to pass more general legislation than that, some introduced in Congress by Senators Sanders and Warren and others, are good. But they don't go far enough. We need to really gut the Taft–Hartley revisions to the National Labor Relations Act, and we need to substitute rules of statute for some of the decisions of the National Labor Relations Board in such areas as striker replacement, joint employment, and some other areas.

So my response to "What is to be done?" is to work to change the laws, and to change them drastically. I invite you to join me in working towards this. I did present a paper on the subject at last year's meeting, and we could use some help in thinking about this, and really seriously considering major changes in the law.

Thank you again for this award. I truly do appreciate it from the bottom of my heart. Thank you.

VIII. LERA Annual Reports

LERA Executive Board Meeting Minutes
Sunday, January 6, 2019, 10:15 a.m. ET
Hyatt Regency Atlanta, Grand Hall East B, Atlanta, GA

Call to order—The meeting was called to order at 10:33 a.m. by Kris Rondeau, President. Present at the meeting were officers Kris Rondeau (President), Dennis Dabney (President Elect), Adrienne Eaton (President Elect Elect), Ryan Lamare (Secretary-Treasurer), Bill Canak (NCAC Chair), and Ariel Avgar (Editor-in-Chief). Board Members in attendance were Matthew Bodah, John Budd (RVP Mid), Bill Dirksen, Cyndi Furseth (Membership Committee Chair), Brad Markell, Dan Marschall, Jim Pruitt (Development Committee Co-Chair), David Weil, and Jeff Wheeler. Aaron Sojourner, Committee Chair for the LERA@ASSA Program Committee, was also present. LERA staff attending the meeting were Emily Smith and Bernadette Tiemann.

Reports

Approval of the Minutes—The motion to approve the minutes passed unanimously from the board meeting in June 2018 and the General Membership Meeting in June 2018; Brad Markell motioned and Matt Bodah seconded.

LERA@ASSA 2019 Program Committee Report—Aaron Sojourner reported. LERA is one of the founding organizations with the ASSA and as such we have a disproportionately large number of sessions allocated to us at the ASSA meeting, and this is an important advantage to maintain open dialog with other economists and policy-makers that meet at this conference. Currently, the ASSA uses a four-year average of median attendance numbers to re-allocate sessions, so attendance is important to maintaining our session allocation at its current level. The program committee is working to improve marketing of our sessions to achieve good attendance, and is also focused on planning excellent sessions, with an array of diverse speakers and research represented. We received $4,500 in sponsorship to offset the cost of the reception and 59 members joined or renewed in conjunction with presenting on this program (45 of these people joined as new members). Bill Canak reminded staff to be sure to send information on new members joining from this program so that they can be encouraged to join their local chapters as well.

Finance Report—Ryan Lamare reported that 2018 has been LERA's strongest year financially in recent times. Our net income was $62,000, which is our third year in row of reporting a net gain. Gains from the last three years have been directed towards refilling the organization's reserves that had been drawn down in prior years.

Dues are higher than projected for 2018; we are up over 100 members from last year, although library subscribers are down. Meeting income is also higher than projected for 2018, partly due to another conference held in conjunction with our own sponsored by the LERA Health Care Industry Council, and other special events that were sponsored, but held at cost. The special events were a positive experience for members but did not add to expenses passed through to meeting registrants, as they were sponsored events. There was a cost savings from *Perspectives on Work*; this was because production management was brought in-house and handled by Bernadette Tiemann on staff.

The 2019 budget has been built with conservative estimates on where we expect our income and expenses to be, but we do expect to be in the black for 2019 as well. We may be seeing a degree of leveling off for 2019 in terms of joins and renewals, but hope to maintain and slightly grow in the future. Ryan Lamare presented the 2017 findings of the auditors for discussion, which had normal findings. The auditors have suggested that we adopt a more traditional accounting methods of reporting our restricted assets so that the balance sheet has a better visualization of what assets are available to the organization. The board will review the various restricted funds at the next board meeting in June, as well as the allied agency agreement with the University of Illinois. Ryan Lamare called for a closed executive session.

After the full meeting re-convened, it was disclosed that the following were approved unanimously by the board: promotion and pay increase for Bernadette Tiemann, pay increase for Emily Smith, and an approval of general salary program as appropriate from the University.

Membership Committee Report—Cyndi Furseth reported that we have had an increase in our member numbers of about 100 people this year, in addition to the increase last year of about 100 members. We have seen a slow-down in the last quarter of 2018 so we may or may not see this type of increase next year. If board members have ideas about membership promotions for 2019 and beyond, we are open to ideas. Apprentices were added as a new category last year, and we currently have 7 apprentice members. Jeff Wheeler suggests that bringing in students as speakers to help bring them into this professional atmosphere. Recruiting young members to help direct the social media of the organization is also helpful. Some chapters allocate 20% of regular member dues to support young members membership and registration to the conference meeting. Having students on committees would be quite helpful, including the membership committee. Let's attempt to get every committee involved with a student. One of the positive membership outcomes in 2019 is the linkage between LRAN and the LERA. LRAN will be holding their conference in Cleveland the day before the LERA 71st Annual Meeting, and Erin Johannson was also just elected to the LERA board. It might be beneficial to develop a strategy to approach LRAN members and help them join the LERA.

Nominating Committee Report—Kris Rondeau reports that the nominating committee has been formulated and requests input from the board of directors. This group of individuals is designed to be independent of the board. David Weil suggests that we invite Aaron Sojourner to be invited to also act on the nominating committee 2019. We will also invite a member or officer of the student LERA chapter at Rutgers University to act on this committee.

In the last election, the following new board members and officers were elected: Adrienne Eaton, (President Elect 2019-20); and board members: Kati Griffith (Academic, 2019-2022); Jake Rosenfeld (Academic, 2019-2022); Heather Boushey (Neutral, 2019-2022); Erin Johnsson (Labor, 2019-2022), Daniel Altchek (Management, 2019-2022), and Robert Chiaravalli (RVP Mid, 2019-2022).

LERA 71st Annual Meeting Program Committee Report—Dennis Dabney and Harry Katz are serving as the Program Committee Co-Chairs for the LERA 71st Annual Meeting in Cleveland, Ohio. Dennis Dabney reported that the committee process of proposal selection was efficient this year, and the Cleveland program has been scheduled and is ready to go. The contract has been signed for the Portland meeting in 2020, with a food and beverage commitment of $30,000 which is manageable and a room block of about 650 room nights, with an 80% attrition rate to meet in order to receive all our concessions in the contract. There are no planned increases in registration pricing for 2019, as the LERA annual meetings have been making ends meet for the last three years. The Labor Research Action Network (LRAN) is planning to conduct their annual meeting on Wednesday the day before the LERA Annual Meeting, also in Cleveland, to coordinate with us.

Editorial Committee Report—Ariel Avgar, LERA editor-in-chief since 2015, reported that the LERA 2018 Research Volume and the *Perspective on Work* Vol. 22 were both released in late November/early December 2018. Each year, the LERA research volume features scholarly research, examining one topic in-depth, and *Perspectives on Work* focuses on short, magazine-like articles that directly appeal to our practitioner members. This volume of *Perspectives on Work* successfully made strides towards being more inclusive in our range of contributing authors, which was a concern brought up at our last board meeting in Baltimore.

The editorial committee continues to have two LERA research volumes in production at any given point. The 2019 volume is currently being produced on the topic: "Employment and Disability: Issues, Innovations, and Opportunities". It's edited by Susanne Bruyere, Lisa Yang, and Hock Tan, all from Cornell University. The 2020 volume is on "Reimagining the Governance of Work and Employment" and will be edited by Dionne Pohler from the University of Toronto. In June, the editorial committee will solicit proposals for the 2021 volume. The committee is engaged in reviewing the model for soliciting proposals and is open to suggestions on improvement/innovation.

The LERA editorial committee also works in partnership with the *ILR Review* to create a "LERA Best Papers" special section in the *ILR Review* … which may at some point become a special volume. Papers presented at a LERA meeting are eligible to participate in this competition. From the 2016 LERA meetings, the competition received 11 papers, of which 6 were reviewed, and 2 made it into the *ILR Review*. Another paper is likely to be approved from the 2017 conference. We have discussed investing in a scholarly LERA journal of our own, but so far this arrangement is proving to be a productive outlet for scholarly publication for our academic members.

National Chapter Advisory Council Report—William Canak, LERA NCAC Chair, reports that LERA chapters are healthy and engaged, with about 40 chapters. We now have a LERA Virtual Chapter, running through Penn State University, which has been conducting meetings using Zoom technology. Tom Kochan spoke at their inaugural meeting with 29 virtual attendees. We will continue to see how they innovate and supply LERA with new resources. The Maine LERA Chapter has established their own website with good features/functionality. The TERRA chapter has been engaged to collaborate with the Tennessee Labor-Management Foundation (TNLMF), with the goal of elevating their annual conference programming. With TERRA's help, Steven Greenhouse presented at the TNLMF Sept. conference, as did Jack Clarke (NAA). In return for assistance with programming, the TNLMF will help with marketing, logistics, and local arrangements at the TERRA conference.

The NCAC will host a chapter representatives meeting and a chapter administration workshop at the LERA 71st Annual Meeting in Cleveland. The RVP's have been functioning well (John Budd, Jim Pruitt, Michele Hoyman). Robert Chiaravalli has just been elected to succeed John Budd to represent the mid-region chapters. He has been a very engaged chapter member and past board member of LERA.

The NCAC proposes that the LERA invite, each year, all the local LERA chapter presidents to attend the annual LERA meeting, and reimburse them for registration following the event. The chapter could send another officer in lieu of the president should that be expedient for the chapter. The NCAC feels that this could be a good investment in the LERA chapters and would keep open the pathways of communication between LERA chapters and the national organization. Matthew Bodah moved the motion to the floor and it was seconded by Kris Rondeau. The motion passed unanimously with no exceptions:

> *Motion:* To encourage interaction between local LERA chapters and LERA national members, facilitate exchange of communication at all levels of the organization, expose LERA Chapter Presidents to resources available through national LERA, and provide a conduit of information from the LERA National Chapter Advisory Council (NCAC) to LERA chapter officers/members, LERA will invite LERA Chapter Presidents to attend the LERA Annual Meeting and to attend the LERA Chapter Representatives Meeting, held at the annual conference. LERA Chapter officers may vote to send a designated chapter elected officer other than the current Chapter President. After the Chapter President or designated, elected chapter officer attends the LERA annual meeting, including the LERA Chapter Representatives Meeting, LERA's conference registration fees will be refunded after the conference. Only one officer per chapter is eligible for this reimbursement.

Development Committee Report—Jim Pruitt, Development Committee Co-Chair, reported that Marlene Heyser, also Co-Chair of the committee, was unable to attend the board meeting as she is currently recovering in the hospital from a serious accident and head injury. We are hoping for a full recovery for Marlene, and many of our members are checking in on her frequently.

The LERA Annual Fund Drive in 2018 was budgeted at $20,000 but brought in $26,000. We received sustaining sponsorship funds in the amount of $45,000 from both UAW/Ford Motor Co. and Coalition of Kaiser Permanente Unions. These sustaining sponsorships have meant a great deal to the health of this organization and we deeply appreciate this support of our mission.

Our organizational members increased from 46 to 59. The committee continues to encourage organizational membership from multiple organizations including AFSCME, Longshore, UFCW, Teamsters, Steelworkers, Unite Here, among others. The work with trying to get more organizations represented within LERA continues with the development committee for 2019.

Our major university contributors included Cornell University and Rutgers University, and Illinois is considering moving into this category for 2019. Other California university programs are also exploring organizational membership with the LERA.

LERA kicked off a new Major Gift Campaign in late 2018 as well, and we will continue to pursue major gifts from individual contributors as well as final gifts and estate gifts. Jim Pruitt made a personal contribution to the major gift campaign for 2018 and invited all other board members and officers to join him in making personal contributions as well, many of whom have done so.

Next meeting will take place on Friday, June 14, 2019 in Cleveland, Ohio, in conjunction with the LERA 71st Annual Meeting.

Adjournment—The meeting was adjourned at 1:12 p.m. by President Kris Rondeau.

LERA Executive Board Meeting Minutes
Friday, June 14, 2019, 8–10 a.m. ET
Westin Cleveland Downtown, Vanda North/South, Cleveland, OH

Call to order—The meeting was called to order at 8:05 a.m. by Kris Rondeau, President. Present at the meeting were officers Kris Rondeau (President), Dennis Dabney (President Elect), Ryan Lamare (Secretary-Treasurer), Bill Canak (NCAC Chair), and Ariel Avgar (Editor-in-Chief). Board Members in attendance were Sylvia Allegretto, Daniel Altchek, John Ammon, Annette Bernhardt, Matthew Bodah, Ezio Borchini, Robert Chiaravalli, Paul Clark, Virginia Doellgast, Jonathon Donehower, Bill Dirksen, Adrienne Eaton, Michelle Hoyman, Brad Markell, Dan Marschall, David Weil, and Jeff Wheeler. Bonnie Castrey, Media Award Committee Chair; Joel Cutcher-Gershenfeld, Chair of the IC/IS Coordinating Committee; and LuAnn Glaser, V-LERA Chapter President, were also present. LERA staff attending the meeting were Emily Smith and Bernadette Tiemann. Unable to attend were John Budd, Cyndi Furseth, Sheila Mayberry, and Jim Pruitt.

Reports

Approval of the Minutes—The motion to approve the minutes passed unanimously from the board meeting in January 2019.

Harry Katz, past LERA President and future ILERA president, asks for the endorsement of the board to plan an overlapping meeting, with up to two co-terminus program days, between ILERA and LERA in the summer of 2024. ILERA rotates their meetings every three years, and in 2024, it will be organized and held in the United States. Tom Kochan and Janice Bellace have both held World Congress' in the United States in recent times, but at that time LERA didn't have the same meeting infrastructure, so this opportunity wasn't available. We would expect 300 or so attendees, and LERA would expect 400-450 attendees, so jointly this would be a large meeting. LA is one proposed meeting site, with several benefits: the availability of an acceptable airport, a reasonably-priced union hotel, the attractiveness of downtown LA, plus the ability to attract those who might wish to attend from Latin America. A motion to approve this arrangement was motioned by Bill Canak, seconded by Brad Markell, and approved unanimously.

Membership/Finance Report—Ryan Lamare reported that membership numbers have been increasing in recent years and have been stabilizing over the last six months. There is no recommendation for a dues increase. In terms of finances, we have been in the black for the last several years and expect this to continue. We project having a positive balance by the end of the year. A significant reason for this is sponsorship that we have been receiving over the last several years. Our auditors have asked us to change our nomenclature from restricted funds to board designated funds, so that change is reflected on the balance sheet. There are no board designated funds that need to be reviewed at this time. The next review is scheduled for 2020. The financial report was adopted unanimously.

LERA 71st Annual Meeting Program Committee Report—Adrienne Eaton, Program Committee Chair, reports that her theme is set. The dates are June 13-16, 2019 and the day rotation is new: Saturday, Sunday, Monday, and Tuesday rather than our usual Thursday-Sunday range. In 2020, we will be celebrating the 100th anniversary of industrial relations in North America. The program committee especially encourages diversity and inclusion in the upcoming program, and requests that session organizers consider the diversity within their own sessions. The committee will meet later today. The committee report was accepted unanimously.

Editorial Committee Report—Ariel Avgar, Editor-in-Chief, reported on the LERA/ILR Review Paper Competition. Two papers in 2016 were accepted and three in 2017 were accepted. Their goal will be to publish an entire special issue of research papers presented at LERA meetings. The LERA 2019 Research Volume on disability is almost complete. Our 2020 volume was approved last year, and will be edited by Dionne Pohler. The committee is reviewing proposals for the 2021 meeting, and we will have proposals to approve at the January 2020 meeting. The *Perspectives on Work* is on track and will deliver an issue in 2019 on the inclusive workplace. The editorial committee report was approved unanimously.

Development Committee Report—Harry Katz reported on behalf of Jim Pruitt. The Development Committee met yesterday with Jim Pruitt. In 2018, the contributions and sponsorships received was $103,000, which is the largest amount received in the last ten years. A major component was major sustaining sponsorships of $45,000, and this year we have received $50,000 from both UAW/Ford Motor Company and Kaiser Permanente/KP Coalition of Unions. We have three major appeals scheduled for 2019 that have yet to be mailed to members (major gift/organizational member/annual fund drive). We have begun to ask people to consider LERA in their estate planning. Last year's ask resulted in a pledge of $100,000 from Gladys Gruenberg, and we have a past pledge from a LERA member of $100,000 as well. Both donors intend those funds to be invested in a quasi-endowment. The second appeal will be for new and renewing organizational members and the third appeal will be the Annual Fund Drive. We are considering adding into those appeals that our 75th anniversary is coming up and we are establishing a 75th anniversary fund. We are expanding the development committee, and we have added a student member, and we will also seek other mid-career members. Harry reports that Jim Pruitt pledges to extend next year's fundraising to more organizations, and that he encourages all board members to consider a personal contribution to LERA based on their own individual abilities. A motion of formal appreciation of the staff was approved unanimously. Bill Canak remarks that Jim Pruitt has been remarkable in his capacity of both DC Chair and as the East Regional Vice President.

Industry Council/Interest Section Coordinating Council Report—Joel Cutcher-Gershenfeld reports. Four of the industry councils asked for business meetings on the 2019 annual meeting program, and they and others have content on the program as well. We had quite a bit of interest at our meeting yesterday to launch a sports, entertainment, and music industry council; they have a great deal of unionized activity in that space. Also, the auto industry council is revitalizing. The dispute resolution interest section, with over 200 members, and the labor and employment law section, with over 200 members, are considering merging. One has active leadership and the other does not. None of the sections have ever had charters, so before they can be merged, they will consider drafting a charter. The DRIS will draft a charter, and then invite the LEL section to join, or invite them to create their own charter. We expect this to take the next year to conduct these activities. We need to re-confirm the leadership and activity for all the interest sections, routinely. Some are quite active and vibrant, but others have become dormant. The idea was proposed to the interest sections to discuss projections in their field for the *Perspectives on Work* magazine.

Nominating Committee Report—Virginia Doellgast presented the slate of candidates for the 2019 LERA election. Wilma Liebman has been asked to run for the next LERA President and will accept if asked. Brad Markell requests that we remove Ana Avendano from the union seat in the slate as she doesn't strictly represent the labor perspective. The slate, as amended, has now been adopted by the board unanimously. Michele Hoyman moved; David Weil seconded the motion.

National Chapter Advisory Council Report—Bill Canak reported on the 2019 chapter awards. The NCAC is working with the Oregon chapter to develop the second LERA student chapter. This is the first year that LERA is offering reimbursement to chapter presidents to attend the LERA annual meeting. The NCAC will also conduct a chapter workshop on Saturday, highlighting how chapters are using technology. Jim Pruitt, John Budd, and Michelle Hoyman are the current Regional Vice Presidents who have begun the work of linking chapters with LERA. This year we will be electing a new East Region Vice President to replace Michelle Hoyman. Please join local LERA chapters and attend their meetings; it is required in our bylaws, and would help us connect the two levels of our organization. Jeff Wheeler motioned to accept; Virginia Doellgast seconded, and the motion passed unanimously.

Regional Vice President Report—Michele Hoyman reported that after making contact with 23 chapters in the east region, it can be said that chapters did not object to the changes in bylaws. Michele will hand over the east region to the next East RVP. Bill Canak motioned to accept the report; it was seconded and the motion approved unanimously.

New Business

Virtual LERA Chapter Report—LuAnn Glaser, VLERA Chapter President, reports on the activities since the chapter started business one year ago. The chapter finished its first year with 101 members. The programs that took place over the last year were presentations by Tom Kochan, Paul Clark, and Eileen

Hoffman, and four more are scheduled for 2019. Officers elected are LuAnn Glaser, President; Michael Wasser, VP; Antone Aboud, Treasurer; and Brandie Laughner, Secretary. The question VLERA is attempting to answer is: how does the virtual chapter provide a sense of community that a local chapter would provide? They have experimented with people introducing themselves on camera, but also do concede that a virtual chapter may not be the same experience. The chapter is discussing whether or not the chapter should have dues; currently there are no expenses other than Chapter to National dues. They are also asking how do we use the digital realm to promote the chapter? One of the issues for 2019 is making members aware of the annual meeting so that VLERA members have the opportunity to attend an in-person meeting if they so choose. Dan Marschall remarks that innovations learned at VLERA can be shared between chapters and used at LERA as well.

LERA Diversity and Inclusion Proposal—Adrienne Eaton reports that other social science organization's policies were reviewed and the proposal has been put together for board review to consider the needs of becoming a more inclusive organization. Brad Markell offered that the AFL-CIO reads a statement at a plenary event at each conference, and they also designate someone who acts as the point of contact for harassment issues. In addition, they make available a business-sized card that displays both the policy and the point of contact should issues arise. LERA might consider asking chapters to adopt a similar policy to our own once it is established. Two motions were approved; one to accept the report, and the second to create a diversity committee to refine our policy and bring it to the next meeting of the board for action.

LERA has engaged an intern to post to our social media outlets. She works five hours a week and distributes all the news items contained in the LERA eBulletin on Twitter, Facebook, and LinkedIn.

Dan Marschall reported that the relationship between LRAN and LERA in 2019 worked well, and that LRAN reported that the arrangement was beneficial. It was noted that some of the sessions were quite similar, which may or may not be a good thing.

Next meeting will take place on Sunday, January 5, 2020 in San Deigo, California in conjunction with the 2020 LERA@ASSA meeting.

Adjournment—The meeting was adjourned at 10:01 a.m. by President Kris Rondeau.

LERA General Membership Meeting and Awards Ceremony 5:30 p.m. to 6:45 p.m., June 15, 2019 Orchid Ballroom, Westin Cleveland Downtown

Call to order. The meeting was called to order at 5:34 p.m. by Kris Rondeau, President. The President stated that LERA meets twice each year, once in June and once in January. Kris introduced Dennis Dabney, President Elect and Adrienne Eaton, the next President Elect. Dennis Dabney is Senior Vice President of Labor Relations at Kaiser Permanente and has 30 years of experience in HR. Adrienne Eaton is the Dean of the School of Management and Labor Relations at Rutgers University.

Committee Reports

Nominating Committee Report. Kris Rondeau reported the results of the last election. Those elected by the membership were President-Elect Adrienne Eaton from Rutgers University (2019-20 President Elect; 2020-21 President; 2021-22 Past President); academic board members were Kati Griffith, Cornell University and Jake Rosenfeld, Washington University at St. Louis; neutral/government/other board member was Heather Boushey, Washington Center for Equitable Growth; the labor board member was Erin Johansson, Jobs with Justice; the management board member was Daniel Altchek, Miles & Stockbridge P.C., and the new RVP Mid Region was Robert Chiaravalli, Strategic Labor and HR LLC. The next election will be held this summer.

Finance and Membership Report. The status of our organization in terms of membership and finance was given by Ryan Lamare. Our membership has been growing over the last several years and has stabilized at around 1,050 members. Strong attendance at the meeting (475) and consistent support from our volunteers and sponsors have resulted in a positive balance for the year. 2018 was the third consecutive year in which we had a positive balance, and 2019 is expected to be as well. Ryan thanked our sponsors, our volunteers, and our staff for these indicators of stability, and invited our members to ask their colleagues to join us.

Development and Contributions Committee Report. Jim Pruitt reports that development was strong in 2018, and in 2019 we the committee will be requesting mid-level contributions from companies and local unions. The committee also appealed to meeting attendees and members to consider making a personal contribution, depending on individual ability to do so. Jim Pruitt reiterated the importance that these personal contributions have had for the organization, and thanked everyone for their support.

Editorial Committee Report. Ariel Avgar, Editor-in-Chief, congratulated the program committee on a terrific conference. He recognized the editorial committee. The *ILR Review*/LERA Best Papers competition, has announced a deadline of July 15 to be considered for this year's competition. The 2019 LERA Research Volume will be published this fall, and has been edited by Susanne Bruyére. Ariel Avgar encouraged members to also ensure that their respective libraries and institutions enjoy access to LERA publications. Dionne Pohler will be editing the 2020 LERA Research Volume. The committee is currently soliciting ideas for the 2021 LERA Research Volume. The *Perspectives on Work* will be published this fall, and will focus on the inclusive work place. The editorial committee thanked both Mike Lillich and Bernadette Tiemann for the wonderful job they are doing with that magazine.

LERA 71st Annual Meeting Program Committee Report. Adrienne Eaton reported that the committee met yesterday to discuss the Portland meeting in 2020. The Call for Sessions is printed on the back of the LERA 70th Annual Meeting program and advertises the November 15, 2019 deadline for proposals. The theme has been selected of sustainability and the world of work, and the meetings will take place next year on Saturday, Sunday, Monday, and Tuesday. The committee is asking that those who are constructing panels to consider diversity of presenters when they are constructing panels. Regionally specific topics are also encouraged, such as marijuana and employment markets, as well as labor and environment or the forestry industry would be of interest. Bruce Kaufman has suggested that 2020 marks the 100th anniversary of the field of IR in North America, and he is invited to submit a session on this topic.

National Chapter Advisory Council Report—Bill Canak invites chapters to submit nominations for chapter awards. He reported that we have a well-functioning and successful student chapter at Rutgers

University, and we are in the process of chartering a second student chapter based at Lewis and Clark University on the West Coast. We look forward to having them join us at the meeting in Portland. Bill thanked and recognized the important role of the Northeast Ohio and Central Ohio LERA Chapters in bringing this event together over the last two years. We have chartered a Virtual LERA (V LERA) chapter based at Penn State University and in their first year they have over 100 members. All their meetings and activities take place online. Bill encourages all chapter members to join LERA and all national members to join local chapters and attend their activities throughout the year. All chapter members are affiliates of the national and get access to our e-Bulletin and online publications, though they cannot vote or receive awards or other LERA member benefits. The NCAC has had some changeover in the makeup of their committee, and they have accepted a number of new members. The bylaws changes in 2016 created three new positions, three regional vice presidents specifically to provide voice for and represent LERA chapters. They each have a vote on the LERA Executive Board. Robert Chiaravalli will succeed John Budd as Mid Regional Vice President, and two nominees will run this summer to succeed Michelle Hoyman as East Regional Vice President. Those candidates will be Beverly Harrison and Tom Wassell.

2019 Awards Ceremony

LERA Media Award—Bill Canak awarded the Ken May Media Award to Moshe Z. Marvit of *The Century Foundation*; Sarah Kessler of *Quartz*; and Dave Jamieson of *The Huffington Post*.

Thomas Kochan and Stephen Sleigh Best Dissertation Award—Bruce Kauffman, BDA Chair, awarded the 2019 Best Dissertation Award to Phillippe Scrimger, Univ. of Montreal for his paper "The Distributive Effects of Trade Unionism: A Look at Income Inequality and Redistribution in Canada's Provinces". Two honorable mentions were awarded in 2019: one to Sean O'Brady, Cornell University for "Negotiating Insecurity? A Comparative Study of Collective Bargaining in Retail Food in Canada, Germany, Sweden and the United States" and one to Paula Marzionna, Cornell University, for "Is This Workplace Bullying? Conflict Management and Workplace Bullying in the Brazilian Banking Sector".

James G. Scoville Best International Paper Award—Ian Greer, BIP Chair, awarded the 2019 Best International Paper to Guglielmo Meardi for "Economic Integration and State Responses: Change in European Industrial Relations since Maastricht".

Susan C. Eaton Research Grant Award—Steve Sleigh conferred the 2019 award for the winning research proposal "Inter-firm contracting, jobs, and inequality in the U.S.: A national, regional, and industry analysis" authored by Jessica Halpern-Finnerty, UC Davis.

ILR Review/LERA Best Papers were presented by Ariel Avgar, Editor-in-Chief and Rose Batt, Editor ILRR. The papers awarded were: Tae-Youn Park, Eun-Suk Lee, and John Budd for the paper "What Do Unions Do for Mothers? Paid Maternity Leave Use and the Multifaceted Roles of Labor Unions"; Rafael Gomez and Danielle Lamb for the paper "Unions and Non-Standard Work: Union Representation and Wage Premiums across Non-Standard Work Arrangements in Canada, 1997–2014"; and Jed DeVaro, Antti Kauhanen, and Nelli Valmari, for the paper "Internal and External Hiring".

John T. Dunlop Outstanding Scholar Awards—Paul Clark presented two 2019 awards. The first award, for outstanding research addressing IR problem of national significance, was given to Maite Tapia, Michigan State University. The second award, for exceptional contributions to international and comparative labor and employment research, was given to J. Adam Cobb, University of Texas at Austin.

The Myron C. Taylor Management Award was presented by Paul Clark. In recognition of outstanding contributions to management in the field of labor relations, this award was presented to William P. Dirksen, Ford Motor Company.

Outstanding Practitioner Award—Paul Clark presented the 2019 award, in recognition of outstanding contributions to practice in the field of labor relations, to Lisa Jordan, United Steelworkers of America and to Lu-Ann Glaser, American Water, Mid-Atlantic Division.

LERA Fellows Awards—Paul Clark presented the 2019 awards. In recognition of outstanding research and practice in the field of labor and employment relations, LERA Fellows were awarded to: John Budd, Joel Cutcher-Gershenfeld, Eileen Appelbaum, David Lewin for their respective academic contributions, and to Jim Pruitt, Owen Herrnstadt, Frances Benson, Bonnie Summers for practitioner contributions to the field of LER.

Susan C. Eaton Outstanding Scholar-Practitioner Award—Paul Clark presented the 2019 award. In recognition of outstanding research and practice emphasizing the value of bringing together the academic and practitioner communities in our field, this award was given to Mark Anner, Pennsylvania State University.

LERA Chapter Star Awards—TERRA (Bill Canak accepting), Long Island LERA (Thomas Wassel accepting), and NE Ohio LERA (Dennis Minni and Greg Szuter accepting) all received Chapter Star Awards for 2019.

New and Other Business

Kris Rondeau, President, announced the dates of the next annual meeting: LERA 72nd Annual Meeting, Portland Hilton, June 13 – 16, 2020 (Sat., Sun., Mon., Tues.), and the LERA@ASSA 2020, San Diego, CA, January 3 - 5, 2020 (with ASSA/AEA). New business and comments were requested from the floor. Canadian Industrial Relations Association announced its calls for papers with a deadline in September for their 2020 regional meeting in Toronto, Canada. The ceremonial gavel was passed to incoming president Dennis Dabney and an award was presented to the outgoing President, Kris Rondeau. The meeting was adjourned by new President Dennis Dabney at 5:45 p.m.

Audit Reports for 2017 and 2018

Labor and Employment Relations Association

Champaign, Illinois

Financial Statements

For the Years Ended December 31, 2018 and 2017

Feller & Kuester CPAs LLP
Certified Public Accountants
806 Parkland Court, Suite 1
Champaign, IL 61821
217-351-3192

TABLE OF CONTENTS

Feller & Kuester CPAs LLP

Tax - Audit - Bookkeeping
806 Parkland Court, Suite #1 - Champaign, Illinois 61821
Phone - (217) 351-3192 Fax - (217) 351-4135 Email - neal@fellerkuester.com

INDEPENDENT AUDITOR'S REPORT

To the Board of Directors of
Labor and Employment Relations Association
Champaign, Illinois

We have audited the accompanying financial statements of the Labor and Employment Relations Association (a nonprofit organization) which comprise the statements of financial position as of December 31, 2018 and 2017, and the related statements of activities, functional expenses, and cash flows for the years then ended, and the related notes to the financial statements.

Management's Responsibility for the Financial Statements
Management is responsible for the preparation and fair presentation of these financial statements in accordance with accounting principles generally accepted in the United States of America; this includes the design, implementation, and maintenance of internal control relevant to the preparation and fair presentation of financial statements that are free from material misstatement, whether due to fraud or error.

Auditor's Responsibility
Our responsibility is to express an opinion on these financial statements based on our audits. We conducted our audits in accordance with auditing standards generally accepted in the United States of America. Those standards require that we plan and perform the audit to obtain reasonable assurance about whether the financial statements are free from material misstatement.

An audit involves performing procedures to obtain audit evidence about the amounts and disclosures in the financial statements. The procedures selected depend on the auditor's judgment, including the assessment of the risks of material misstatement of the financial statements, whether due to fraud or error. In making those risk assessments, the auditor considers internal control relevant to the entity's preparation and fair presentation of the financial statements in order to design audit procedures that are appropriate in the circumstances, but not for the purposes of expressing an opinion on the effectiveness of the entity's internal control. Accordingly, we express no such opinion. An audit also includes evaluating the appropriateness of accounting policies used and the reasonableness of significant accounting estimates made by management, as well as evaluating the overall presentation of the financial statements.

We believe that the audit evidence we have obtained is sufficient and appropriate to provide a basis for our audit opinion.

Opinion

In our opinion, the financial statements referred to above present fairly, in all material respects, the financial position of the Labor and Employment Relations Association as of December 31, 2018 and 2017, and the changes in its net assets and its cash flows for the years then ended in accordance with the accounting principles generally accepted in the United States of America.

Feller & Kuester CPAs LLP

Feller & Kuester CPAs LLP
Champaign, Illinois

May 23, 2019

LABOR AND EMPLOYMENT RELATIONS ASSOCIATION
STATEMENTS OF FINANCIAL POSITION
DECEMBER 31, 2018 AND 2017

	2018	2017
ASSETS		
Current Assets		
Cash and Cash Equivalents	$ 233,798	$ 170,028
Investments	24,663	24,709
Accounts Receivable, Net	583	-
Prepaid Expenses	8,216	8,131
Inventory	6,374	5,445
Accrued Interest and Royalties	2,579	7,764
Total Current Assets	276,213	216,077
Property and Equipment, Net	1,508	2,047
TOTAL ASSETS	$ 277,721	$ 218,124
LIABILITIES AND NET ASSETS		
Current Liabilities		
Accounts Payable	$ 82	$ 583
Funds Held for Other Organizations	13,025	10,411
Accrued Liabilities	30,595	25,170
Dues Collected in Advance	72,754	68,538
Subscriptions Collected in Advance	12,848	10,091
Deferred Chapter and Organization Dues	5,050	9,538
Other Deferred Revenue	13,700	8,000
Total Current Liabilities	148,054	132,331
Net Assets		
Without Donor Restrictions		
Designated by the Board		
Susan C. Eaton Scholar-Practitioner Memorial Fund	26,779	29,779
Kochan-Sleigh Best Dissertation Award Fund	39,791	40,899
Jim Scoville Award Fund	4,539	5,147
John T. Dunlop Public Policy Fund	14,792	18,828
Gladys and Water Gershenfeld Publication Fund	16,988	20,997
Total Designated by the Board	102,889	115,650
Undesignated	26,778	(29,857)
Total Net Assets Without Donor Restrictions	129,667	85,793
TOTAL LIABILITIES AND NET ASSETS	$ 277,721	$ 218,124

See Accompanying Notes.

LABOR AND EMPLOYMENT RELATIONS ASSOCIATION
STATEMENTS OF ACTIVITIES
FOR THE YEARS ENDED DECEMBER 31, 2018 AND 2017

	2018			2017		
	Without Donor Restrictions			Without Donor Restrictions		
	Undesignated	Designated	Total	Undesignated	Designated	Total
Support and Revenues						
Membership Dues	$ 151,608	$ -	$ 151,608	$ 133,391	$ -	$ 133,391
Meeting Income	209,686	-	209,686	146,441	-	146,441
Organization Dues & Sponsorships	58,150	-	58,150	28,700	-	28,700
Subscriptions	16,770	-	16,770	20,651	-	20,651
Chapter Fees	8,338	-	8,338	7,700	-	7,700
Member Contributions	28,321	-	28,321	23,586	1,899	25,485
Royalties	11,718	-	11,718	9,485	-	9,485
Ad Income	7,300	-	7,300	10,100	-	10,100
Publications	34	-	34	1,095	-	1,095
Mailing List Rental	815	-	815	957	-	957
Administrative Fees	448	-	448	256	-	256
Investment Income	-	(46)	(46)	-	913	913
Interest Income	899	-	899	157	427	584
Net Assets Released from Designation	12,715	(12,715)	-	12,401	(12,401)	-
Total Support and Revenues	506,802	(12,761)	494,041	394,920	(9,162)	385,758
Expenses						
Program Services						
General	183,464	-	183,464	178,539	-	178,539
Meetings	153,143	-	153,143	101,036	-	101,036
Publications	35,694	-	35,694	36,811	-	36,811
Supporting Services						
Management and General	62,918	-	62,918	57,029	-	57,029
Membership Development	14,948	-	14,948	9,201	-	9,201
Total Expenses	450,167	-	450,167	382,616	-	382,616
Change in Net Assets	56,635	(12,761)	43,874	12,304	(9,162)	3,142
Net Assets, Beginning of Year	(29,857)	115,650	85,793	(42,161)	124,812	82,651
Net Assets, End of Year	$ 26,778	$ 102,889	$ 129,667	$ (29,857)	$ 115,650	$ 85,793

See Accompanying Notes.

LABOR AND EMPLOYMENT RELATIONS ASSOCIATION
STATEMENT OF FUNCTIONAL EXPENSES
FOR THE YEAR ENDED DECEMBER 31, 2018

		MEETINGS			PUBLICATIONS			SUPPORTING SERVICES		
	General	Annual Conference	ASSA Meeting	Other Meetings	Perspectives	Research Volume	Other Publications	Management & General	Membership Development	Totals
Compensation	$ 121,005	$ -	$ -	$ -	$ -	$ -	$ -	$ 23,354	$ -	$ 144,359
Payroll Taxes & Fringes	46,599	-	-	-	-	-	-	7,978	-	54,577
Contract Labor	15,860	-	-	-	-	-	-	-	-	15,860
Depreciation	-	-	-	-	-	-	-	539	-	539
Insurance	-	-	-	-	-	-	-	849	-	849
Bank and Service Charges/Fulfillment	-	-	-	-	-	-	-	9,073	-	9,073
Promotion	-	9,623	-	-	-	-	-	-	3,951	13,574
Postage/Freight	-	73	57	-	1,762	2,299	99	371	-	4,661
Accounting/Auditing	-	-	-	-	-	-	-	5,036	-	5,036
Printing and Production	-	2,678	95	-	6,541	7,540	459	-	-	17,313
Services	-	5,067	-	-	7,079	8,159	1,382	-	-	21,687
Other Publication Costs	-	-	-	-	285	89	-	-	-	374
Meals/Receptions	-	62,919	4,897	7,409	-	-	-	-	-	75,225
Travel	-	3,871	2,539	8,025	-	-	-	-	-	14,435
Lodging	-	-	-	-	-	-	-	-	-	-
Other Meeting Expenses	-	45,510	280	100	-	-	-	-	-	45,890
Computer Supplies/Services	-	-	-	-	-	-	-	7,346	-	7,346
Website	-	-	-	-	-	-	-	5,733	-	5,733
Office Supplies	-	-	-	-	-	-	-	2,038	-	2,038
Student & Member Awards	-	-	-	-	-	-	-	-	7,032	7,032
Fundraising Expense	-	-	-	-	-	-	-	-	2,521	2,521
Duplicating Expense	-	-	-	-	-	-	-	433	-	433
Telephone & Fax	-	-	-	-	-	-	-	-	-	-
Chapter Expenses	-	-	-	-	-	-	-	-	542	542
Other Committee Expenses	-	-	-	-	-	-	-	-	902	902
Miscellaneous Office	-	-	-	-	-	-	-	168	-	168
Total	$ 183,464	$ 129,741	$ 7,868	$ 15,534	$ 15,667	$ 18,087	$ 1,940	$ 62,918	$ 14,948	$ 450,167

See Accompanying Notes.

LABOR AND EMPLOYMENT RELATIONS ASSOCIATION
STATEMENT OF FUNCTIONAL EXPENSES
FOR THE YEAR ENDED DECEMBER 31, 2017

		MEETINGS			PUBLICATIONS			SUPPORTING SERVICES		
	General	Annual Conference	ASSA Meeting	Other Meetings	Perspectives	Research Volume	Other Publications	Management & General	Membership Development	Totals
Compensation	$ 117,626	$ -	$ -	$ -	$ -	$ -	$ -	$ 18,688	$ -	$ 136,314
Payroll Taxes & Fringes	42,821	-	-	-	-	-	-	6,946	-	49,767
Contract Labor	18,092	-	-	-	-	-	-	-	-	18,092
Depreciation	-	-	-	-	-	-	-	539	-	539
Insurance	-	-	-	-	-	-	-	2,617	-	2,617
Bank and Service Charges/Fulfillment	-	-	-	-	-	-	-	12,145	-	12,145
Promotion	-	3,623	-	-	-	-	-	-	1,156	4,779
Postage/Freight	-	1,127	127	-	1,550	1,610	-	916	-	5,330
Accounting/Auditing	-	-	-	-	-	-	-	4,872	-	4,872
Printing and Production	-	2,014	-	-	7,232	6,441	-	-	-	15,687
Services	-	-	-	-	11,733	6,159	1,557	-	-	19,449
Other Publication Costs	-	-	-	-	257	-	272	-	-	529
Meals/Receptions	-	73,302	5,465	-	-	-	-	-	-	78,767
Travel	-	4,110	455	-	-	-	-	-	-	4,565
Lodging	-	4,896	-	-	-	-	-	-	-	4,896
Other Meeting Expenses	-	3,876	471	1,570	-	-	-	-	-	5,917
Computer Supplies/Services	-	-	-	-	-	-	-	4,434	-	4,434
Website	-	-	-	-	-	-	-	4,569	-	4,569
Office Supplies	-	-	-	-	-	-	-	481	-	481
Student & Member Awards	-	-	-	-	-	-	-	-	6,407	6,407
Fundraising Expense	-	-	-	-	-	-	-	-	871	871
Duplicating Expense	-	-	-	-	-	-	-	476	-	476
Telephone & Fax	-	-	-	-	-	-	-	327	-	327
Chapter Expenses	-	-	-	-	-	-	-	-	363	363
Other Committee Expenses	-	-	-	-	-	-	-	-	404	404
Miscellaneous Office	-	-	-	-	-	-	-	19	-	19
Total	$ 178,539	$ 92,948	$ 6,518	$ 1,570	$ 20,772	$ 14,210	$ 1,829	$ 57,029	$ 9,201	$ 382,616

See Accompanying Notes.

LABOR AND EMPLOYMENT RELATIONS ASSOCIATION
STATEMENTS OF CASH FLOWS
FOR THE YEARS ENDED DECEMBER 31, 2018 AND 2017

	2018	2017
CASH FLOWS FROM OPERATING ACTIVITIES		
Change in Net Assets	$ 43,874	$ 3,142
Adjustments to Reconcile Change in Net Assets to Net Cash Provided by (Used in) Operating Activities		
Depreciation Expense	539	539
Net Unrealized (Gains) Losses on Investments	736	(252)
(Increase) Decrease in Operating Assets		
Accounts Receivable, Net	(583)	-
Prepaid Expenses	(85)	2,841
Inventory	(929)	1,177
Accrued Interest and Royalties	5,185	(7,764)
Increase (Decrease) in Operating Liabilities		
Accounts Payable	(501)	(14,838)
Funds Held for Other Organizations	2,614	3,806
Accrued Liabilities	5,425	4,746
Dues Collected in Advance	4,216	7,125
Subscriptions Collected in Advance	2,757	(3,493)
Deferred Chapter Dues	(4,488)	1,738
Other Deferred Revenue	5,700	6,500
Net Cash Provided by (Used in) Operating Activities	64,460	5,267
CASH FLOWS FROM INVESTMENT ACTIVITIES		
Reinvested Interest, Dividends, and Capital Gains	(690)	(661)
Purchases of Property and Equipment	-	(2,310)
Net Cash Provided by (Used in) Investment Activities	(690)	(2,971)
CASH FLOWS FROM FINANCING ACTIVITIES		
None	-	-
Net Cash Provided by (Used in) Financing Activities	-	-
Net Increase (Decrease) in Cash and Cash Equivalents	63,770	2,296
Beginning Cash and Cash Equivalents	170,028	167,732
Ending Cash and Cash Equivalents	$ 233,798	$ 170,028

See Accompanying Notes.

NOTE 1 – NATURE OF ACTIVITIES AND SIGNIFICANT ACCOUNTING POLICIES

Organization and Nature of Activities

The Labor and Employment Relations Association (the Organization) was founded in 1947 to encourage research in all aspects of the field of labor, employment, and the workplace. It is a nonprofit scholarly association of academic, labor, business, and neutral communities committed to the full discussion and exchange of ideas between and among its broad constituencies through meetings, publications, and its various electronic listservs and websites. The Organization's National Office is located in Champaign, Illinois and serves the association by planning conferences and meetings and publishing the various researches of its members. The main sources of support and revenues for the Organization are contributions, membership dues and subscriptions, and meeting income.

Subsequent Events

Subsequent events have been evaluated through May 23, 2019, which is the date the financial statements were available to be issued.

Accrual Basis of Reporting

The Organization has chosen to report on the accrual basis of accounting. Accordingly, revenue is recognized when earned and expenses are recognized when incurred in conformity with accounting principles generally accepted in the United States of America (GAAP). The financial statements are presented in accordance with Financial Accounting Standards Board (FASB) Accounting Standards Codification (ASC) 958 dated August 2016, and the provisions of the American Institute of Certified Public Accountants (AICPA) "Audit and Accounting Guide for Not-for-Profit Organizations" (the "Guide").

Estimates

The preparation of financial statements in conformity with GAAP requires management to make estimates and assumptions that affect certain reported amounts and disclosures. Accordingly, actual results could differ from those estimates.

Cash and Cash Equivalents

For purposes of reporting cash flows, cash and cash equivalents include all cash and highly liquid investments acquired with an original maturity date of three months or less. Since the penalties of converting certificate of deposits to cash is insignificant, all certificate of deposits have been included with cash and cash equivalents. As of December 31, 2018 and 2017, $78,226 and $90,941 of cash and cash equivalents are designated by the board and are subject to board-imposed stipulations

Investments

Investments consist of intermediate term bond index funds which are carried at the fair value of the underlying assets. Net appreciation (depreciation) in the fair value of investments, which consists of the realized gains or losses and the unrealized appreciation (depreciation) on those investments, is presented in the statement of activities in accordance with donor restrictions as investment income. Investment income is presented net of investment fees. The average cost method is primarily used to determine the basis for computing realized gains or losses. All investments are held as designated by the board and are subject to board-imposed stipulations.

Accounts Receivable

Accounts receivable are recorded primarily for outstanding invoices for membership dues and subscriptions. An allowance for doubtful accounts is based on an analysis of expected collection rates determined from experience. The Organization had an allowance for doubtful accounts of $0 and $0 at December 31, 2018 and 2017, respectively.

Inventory

The Organization's inventory of directories, research volumes, proceedings, and perspective magazines is carried at the lower of cost and market value. Cost is determined on the basis of first in – first out.

Property and Equipment

Property and equipment expenditures in excess of $500 are capitalized at cost. Donated property and equipment are capitalized at estimated cost or fair market value at the time of donation. Depreciation of the assets is computed using the straight-line method over their estimated useful lives. The range of estimated useful lives by type of asset is as follows:

Furniture and Equipment	5-7 years

Net Assets

Net assets of the Organization and changes therein are classified and reported as follows:

Net Assets without Donor Restrictions - Net assets that are not subject to donor-imposed restrictions and may be expended for any purpose in performing the primary objectives of the Organization. The Organization's board may designate assets without restrictions for specific operational purposes from time to time.

Net Assets with Donor Restrictions - Net assets subject to stipulations imposed by donors, and grantors. Some donor restrictions are temporary in nature; those restrictions will be met by actions of the Organization or by the passage of time. Other donor restrictions are perpetual in nature, where by the donor has stipulated the funds be maintained in perpetuity. The Organization had no net assets with donor restrictions as of December 31, 2018 and 2017.

Contributions

Unconditional contributions are recognized when pledged and recorded as net assets without donor restrictions or net assets with donor restrictions, depending on the existence and/or nature of any donor-imposed restrictions. Conditional promises to give are recognized when the conditions on which they depend are substantially met. Gifts of cash and other assets are reported with donor restricted support if they are received with donor stipulations that limit the use of the donated assets.

When a restriction expires, that is, when a stipulated time restriction ends or a purpose restriction is accomplished, net assets with donor restrictions are reclassified to net assets without donor restrictions and reported in the statement of activities as net assets released from restrictions. Donor-restricted contributions whose restrictions are met in the same reporting period are reported as net assets without donor restriction support. Contributions restricted for the acquisition of land, buildings, and equipment are reported as net assets without donor restriction upon acquisition of the assets and the assets are placed in service.

Contributed Services and Goods

Contributed services are reported as contribution revenue and as assets or expenses only if the services create or enhance a non-financial asset (for example property and equipment) or:

- Would typically need to be purchased by the Organization if the services had not been provided by contribution.
- Require specialized skills.
- Are provided by individuals with those skills (such as accounting, financial, construction, educational, electrical, legal, medical, and other services provided by accountants, investments advisers, contractors, teachers, electricians, lawyers, doctors, and other professional and craftspeople).

For the years ended December 31, 2018 and 2017, the value of contributed services meeting the defined requirements for recognition in the financial statements as outlined above were not material and have not been recorded on the financial statements. The Organization does receive free office space from the University of Illinois. However, the value for the use of this office space has not been reported on the financial statements.

Income Tax Status

The Organization is a nonprofit association that is exempt from federal income tax under Section 501(c)(3) of the Internal Revenue Code. In addition, the Internal Revenue Service has determined that the Organization is not a private foundation as defined in Section 509(a)(1) and Section 170(b)(1)(A)(vi) of the Code.

The Organization has evaluated its exposure resulting from uncertain income tax position and determined the exposure is not material to the financial statements. In addition, the Organization is not aware of any tax position for which a significant change is reasonably possible within the next 12 months. Therefore, these financial statements do not include a liability for uncertain tax positions. Upon recognition of a liability for an uncertain tax position, the Organization would recognize interest expense and penalties in operating expenses.

The Organization files information tax returns in the U.S. federal jurisdiction and the state of Illinois. Its federal and Illinois information tax returns prior to fiscal year 2015 are closed. The Organization does not have any tax returns currently under examination by either the Internal Revenue Service (IRS) or any U.S. state jurisdiction.

Membership Dues and Advance Subscriptions Collected

Membership dues and subscriptions are assessed and recognized as revenue based on the life of the dues or subscription.

Functional Allocation of Expenses

The costs of providing the various programs and other activities have been summarized on a functional basis in the statements of activities. The statements of functional expenses present the natural classification detail of expenses by function. Accordingly, certain costs have been allocated among the programs and supporting services benefited.

Expenses which are easily and directly associated with a particular program or supporting service are charged directly to that functional area. Compensation and payroll taxes & fringes have been allocated on the basis of estimates of time and effort.

New Accounting Pronouncement

On August 18, 2016, FASB issued ASU 2016-14, Not-for-Profit Entities (Topic 958) – Presentation of Financial Statements of Not-for-Profit Entities. The update addresses the complexity and understandability of net asset classification, deficiencies in information about liquidity and availability of resources, and the lack of consistency in the type of information provided about expenses and investment return. The Organization has adjusted the presentation of these statements accordingly. The ASU has been applied retrospectively to all periods presented.

NOTE 2 – ARRANGEMENTS WITH THE UNIVERSITY OF ILLINOIS

The Organization moved its offices to the University of Illinois at the end of 1999. Under an arrangement with the University, the employees of the Organization are employed by the University. The employees' pension and benefits are part of the University's plans. The Organization then reimburses the University monthly for the cost of its employees.

The University of Illinois holds some cash for the Organization. These "claim on cash" balances were $1,878 and $24,352 as of December 31, 2018 and 2017, respectively. These balances have been included with cash and cash equivalents.

NOTE 3 – SERVICING ARRANGEMENTS TO ADMINISTER BUSINESS

On January 3, 2003, the Organization entered into a servicing arrangement to administer the business of the University Council of Industrial Relations and Human Resources Programs (UCIRHRP). These servicing duties include dues notification and collection, annual meeting arrangement and report preparation, and maintaining a data base and network communications for this separate organization. A separate financial statement is maintained for this company, but UCIRHRP's cash is maintained in the Organization's general bank account. The amount of cash belonging to UCIRHRP in the general account is included on the Organization's balance sheet as Funds Held for Other Organizations. UCIRHRP's cash balance in the Organization's custody was $7,380 at December 31, 2018 and $7,160 at December 31, 2017. The Organization is allowed to keep 20% of dues collected each year as an administrative fee.

The Organization entered into an agreement to administer the PhD Student Consortium. The Organization's duties include collecting donations and issuing stipends for eligible students. The amount of cash belonging to the PhD Student Consortium is included on the Organization's balance sheet as Funds Held for Other Organizations. PhD Student Consortium's cash balance in the Organization's custody was $5,645 at December 31, 2018 and $3,251 at December 31, 2017.

NOTE 4 – CONCENTRATIONS OF CREDIT RISK

Financial instruments that potentially subject the Organization to credit risk consist principally of checking accounts, money markets accounts, and certificates of deposits at financial institutions. However, management continuously monitors the Organization's balances at financial institutions. Deposits held at any financial institutions were fully insured by the Federal Deposit Insurance Corporation (FDIC) at December 31, 2018 or 2017. The "claims on cash" held with the University of Illinois as stated in Note 2 is not insured since it is not held in a financial institution. In addition, the Organization had undeposited funds of $5,441 and $0 at December 31, 2018 and 2017, respectively. Since these funds were not yet deposited with a financial institution it was covered by FDIC.

NOTE 5 – BOARD DESIGNATED NET ASSETS

Susan C. Eaton Scholar-Practitioner Memorial Fund

The Organization set up a memorial fund in honor of an author of a 1998 "Perspectives on Work" article who died on December 30, 2003. At its June 1, 2004 meeting, the Organization approved the establishment of an annual Susan C. Eaton Scholar-Practitioner award and grant to be paid to one or more qualified scholar researchers in even-numbered years or practitioners in odd-numbered years doing research in the labor and employment relations or related field. The Organization's Executive Board directed that 10% of any gifts received each year are undesignated and can be used for administrative expenses. As of December 31, 2018 and 2017, these designated funds totaled $26,779 and $29,779, respectively.

Kochan-Sleigh Best Dissertation Award Fund

On March 27, 2006, the Organization set up a designated fund with contributions from a member and matching contributions from General Electric. The fund is designated for a minimum of fifteen years and may be used to pay for the $1,000 annual best dissertation award and plaque. After the fifteen-year period, the fund may be continued or the amount remaining in the fund may be undesignated and become available to the Organization for unrestricted purposes. The Organization's Executive Board directed that 10% of any gifts received each year are undesignated and can be used for administrative expenses. As of December 31, 2018 and 2017, these designated funds totaled $39,791 and $40,899, respectively.

Jim Scoville Award Fund

On January 9, 2009, the Organization set up a designated fund with a $10,000 contribution from the University of Minnesota's Industrial Relations Center to honor a member and retiring professor. The fund was established to pay an annual award of $500 and a plaque for best paper on international and comparative employment issues. The Organization's Executive Board directed that 10% of any gifts received each year are undesignated and can be used for administrative expenses. As of December 31, 2018 and 2017, these designated funds totaled $4,539 and $5,147, respectively.

John T. Dunlop Public Policy Fund

On May 8, 2010, the Organization established a designated fund to subsidize a named John T. Dunlop Public Policy Session at future National Policy Forums and/or at the Organization annual meetings. The Organization's Executive Board directed that 10% of any gifts received each year are undesignated and can be used for administrative expenses. This fund will be so named for five or ten years, at which time the Organization's Executive Board will review the fund to determine whether to continue or modify the fund. As of December 31, 2018 and 2017, these designated funds totaled $14,792 and $18,828, respectively.

Gladys and Walter Gershenfeld Publication Fund

On May 8, 2010, the Organization established a designated fund to honor the memory of long-time members and past presidents Walter and Gladys Gershenfeld for the purpose of supporting the Organization's electronic and print publications. The fund will be so named for ten years, at which time the Organization will review the fund's purpose and uses, and whether to continue or modify the fund. The Organization's Executive Board directed that 10% of any gifts received each year are undesignated and can be used for administrative expenses. As of December 31, 2018 and 2017, these designated funds totaled $16,988 and $20,997, respectively.

NOTE 6 – INVESTMENTS

FASB Codification 820, Fair Value Measurements, establishes a framework for measuring fair value. That framework provides a fair value hierarchy that prioritizes the inputs to valuation techniques used to measure fair value. The hierarch gives the highest priority to unadjusted quoted prices in active markets for identical assets or liabilities (Level 1 measurements) and the lowest priority to unobservable inputs (Level 3 measurements). The three levels of the fair value hierarchy under FASB Codification 820 are described below:

> Level 1 – Inputs to the valuation methodology are based on unadjusted quoted prices for identical assets or liabilities in active markets that the Organization has the ability to access.
>
> Level 2 – Inputs to the valuation methodology include quoted prices for similar assets or liabilities in active markets, quoted prices for identical or similar assets or liabilities in inactive markets, inputs other than quoted prices that are observable for the asset or liability, and inputs that are derived principally from or corroborated by observable market data by correlation or other means. If the asset or liability has a specified (contractual) term, the Level 2 input must be observable for substantially the full term of the asset or liability.
>
> Level 3 – Inputs to the valuation methodology are unobservable and significant to the fair value measurement.

The asset's and liability's fair value measurement level within the fair value hierarchy is based on the lowest level of any input that is significant to the fair value measurement. Valuation techniques used need to maximize the use of observable inputs and minimize the use of unobservable inputs.

The following is a description of the valuation methodologies used for assets measured at fair value. There have been no changes in the methodologies used at December 31, 2018 and 2017.

> *Intermediate Term Bond Funds:* Valued at the NAV of shares held by the Organization at year-end based on readily determinable fair values, which are published daily and are the basis for current transactions.

The preceding methods described may produce a fair value calculation that may not be indicative of net realizable value or reflective of future fair values. Furthermore, although the Organization believes its valuation methods are appropriate and consistent with other market participants, the use of different methodologies or assumptions to determine the fair value of certain financial instruments could result in different fair value measurements at the reporting date.

The following table sets forth by level, within the fair value hierarchy, the Organization's assets at fair value as of December 31, 2018 and 2017:

Assets at Fair Value as of December 31, 2018

	Level 1	Level 2	Level 3	Total
Intermediate Term Bond Funds	$24,663	-	-	$24,663

Assets at Fair Value as of December 31, 2017

	Level 1	Level 2	Level 3	Total
Intermediate Term Bond Funds	$24,709	-	-	$24,709

Net investment earnings for the years ended December 31, 2018 and 2017 are summarized as follows:

	2018	2017
Dividends	$ 690	$ 637
Capital Gains Distributions	-	24
Net Unrealized Gains (Losses)	(736)	252
Total Investment Income	$ (46)	$ 913

This investment income, including unrealized gains and losses, are being reported on the statement of activity as investment income.

NOTE 7 – PROPERTY AND EQUIPMENT

As of December 31, 2018, Property and Equipment consists of:

Furniture and Equipment	$ 15,262
Less: Accumulated Depreciation	(13,754)
Property and Equipment, Net	$ 1,508

As of December 31, 2017, Property and Equipment consists of:

Furniture and Equipment	$ 15,262
Less: Accumulated Depreciation	(13,215)
Property and Equipment, Net	$ 2,047

Depreciation expense for the years ended December 31, 2018 and 2017 was $539 and $539, respectively.

NOTE 8 – ACCRUED EXPENSES

As of December 31, 2018, Accrued Expenses consist of:

Accrued Payroll	$ 5,861
Accrued Benefits/Taxes	3,977
Accrued Vacation	20,757
Total	$ 30,595

As of December 31, 2017, Accrued Expenses consist of:

Accrued Payroll	$ 5,202
Accrued Benefits/Taxes	3,207
Accrued Vacation	16,761
Total	$ 25,170

NOTE 9 – LIQUIDITY AND AVAILABILITY OF FINANCIAL ASSETS

The following reflects the Organization's financial assets as of the statements of financial position date, reduced by amounts not available for general use because of contractual or donor-imposed restrictions within one year of the statement of financial position date. Amounts available include donor restricted amounts that are available for general expenditure in the following year. Amounts not available include amounts set aside for the items described in Note 5 that could be drawn upon if the Board of Directors approves that action.

	2018	2017
Fiscal Assets at Year-End		
Cash and Cash Equivalents	$ 233,798	$ 170,028
Investments	24,663	24,709
Accounts Receivable, Net	583	-
Prepaid Expenses	8,216	8,131
Inventory	6,374	5,445
Accrued Interest and Royalties	2,579	7,764
Less Contractual or Donor-Imposed Restrictions		
Board Designated Funds	(102,889)	(115,650)
Financial Assets Available to Meet Cash Needs for General Expenditure Within One Year	$ 173,324	$ 100,427

NOTE 10 – COMMITMENTS

On February 26, 2013, the Organization signed a twelve-month contract from March 1, 2013 to February 28, 2014 with a company to provide services to manage the EPRN website and provide content for the Organization's website. The contract states that the Organization will pay this company $2,400 per month for a total of $28,800. The Agreement automatically renews for consecutive twelve-month terms, unless the Agreement is terminated by one or both of the parties as set forth herein. Beginning the second cycle and continuing until termination or renegotiation, the Organization will increase the base rate every year by two percent (2%) over the previous year's base rate. For example: $2,448 per month for twelve months starting March 1, 2014; $2,497 per month for twelve months starting March 1, 2015; $2,546 per month for twelve months starting March 1, 2016; and so on.

On January 18, 2017, the Organization signed a new modified twelve-month contract from March 1, 2017 to February 28, 2018 with the same company to provide services to provide content for the Organization's website. The contract states that the Organization will pay this company $1,300 per month for a total of $15,600. The Agreement automatically renews for consecutive twelve-month terms, unless the Agreement is terminated by one or both of the parties as set forth herein. Beginning the second cycle and continuing until termination or renegotiation, the Organization will increase the base rate every year by two percent (2%) over the previous year's base rate. For example: $1,326 per month for twelve months starting March 1, 2018; $1,353 per month for twelve months starting March 1, 2019; and so on.

The Organization has signed contracts with hotels in Baltimore, Cleveland, and Portland for the 2018, 2019, and 2020 Annual Meetings. If these contracts were cancelled at December 31, 2018 and 2017, the Organization would have owed $94,387 and $113,655, respectively, to the Baltimore, Cleveland, and Portland hotels.

NOTE 11 – PENSION PLAN AND RETIREMENT CONTRIBUTIONS

The University of Illinois (the University) contributes to the State Universities Retirement System of Illinois (SURS), a cost-sharing multiple-employer defined benefit plan with a special funding situation whereby the State of Illinois (the State) makes substantially all actuarially determined required contributions on behalf of the participating employers. SURS was established July 21, 1941 to provide retirement annuities and other benefits for staff members and employees of state universities, certain affiliated organizations, and certain other state educational and scientific agencies and for survivors, dependents, and other beneficiaries of such employees. SURS is considered a component unit of the State of Illinois' financial reporting entity and is included in the State's financial reports as a pension trust fund. SURS is governed by Section 5/15, Chapter 40 of the *Illinois Compiled Statutes*. SURS issues a publicly available financial report that includes financial statements and required supplementary information. That report may be obtained by accessing the website at www.SURS.org.

Benefits Provided. A traditional benefit plan was established in 1941. Public Act 90-0448 enacted effective January 1, 1998, established an alternative defined benefit program known as the portable benefit package. The traditional and portable plan Tier 1 refers to members that began participation prior to January 1, 2011. Public Act 96-0889 revised the traditional and portable benefit plans for members who begin participation on or after January 1, 2011, and who do not have other eligible Illinois reciprocal system services. The revised plan is referred to as Tier 2. New employees are allowed 6 months after their date of hire to make an irrevocable election. A summary of the benefit provisions can be found in the SURS' comprehensive annual financial report (CAFR) Notes to the Financial Statements.

Contributions. The State of Illinois is primarily responsible for funding SURS on behalf of the individual employers at an actuarially determined amount. Public Act 88-0593 provides a Statutory Funding Plan consisting of two parts: (i) a ramp-up period from 1996 to 2010 and (ii) a period of contributions equal to a level percentage of the payroll of active members of the System to reach 90 percent of the total Actuarial Accrued Liability by the end of fiscal year 2045. Employer contributions from "trust, federal, and other funds" are provided under Section 15-155(b) of the Illinois Pension Code and require employers to pay contributions which are sufficient to cover the accruing normal costs on behalf of applicable employees. The employer normal cost was 12.46 percent of employee payroll during the period of July 1, 2017 to June 30,

2018. The employer normal cost was 12.29 percent of employee payroll during the period of July 1, 2018 to June 30, 2019. The normal cost is equal to the value of current year's pension benefit and does not include any allocation for the past unfunded liability or interest on the unfunded liability. Plan members are required to contribute 8.0 percent of their annual covered salary. The contribution requirements of plan members and employers are established and may be amended by the Illinois General Assembly. During the years ended December 31, 2018, 2017 and 2016, the Organization reimbursed the University of Illinois $16,958, $14,764, and $10,388, respectively, for the employer's payments to SURS for eligible employees.

In addition to providing pension benefits, the State of Illinois provides certain health, dental and life insurance benefits to annuitants. This includes annuitants of the Organization. Substantially all State employees, including the Origination's employees, may become eligible for postemployment benefits if they eventually become annuitants. Health and dental benefits include basic benefits for annuitants under the State's self-insurance plan and insurance contracts currently in force. Life insurance benefits for annuitants under age 60 are equal to their annual salary at the time of retirement; life insurance benefits for annuitants age 60 or older are limited to $5,000 per annuitant. Currently, the State does not segregate payments made to annuitants from those made to current employees for health, dental and life insurance benefits. These costs are funded by the State and are not an obligation of the Organization.

Employees of the Organization may also elect to participate in several tax deferred annuity plans and defined contribution plans. These are single employer plans under which benefits are provided to participating employees through contracts issued to each individual. Participation and the level of employee contributions are voluntary. The Organization is not required to make contributions.

LERA gratefully acknowledges the continuing support of its 2018–19 Major Sponsors and Organizational Members

Major Sustaining Sponsors

Kaiser Permanente and the Labor Management Partnership
UAW/Ford National Programs

Major University Sponsors

Cornell University, School of Industrial and Labor Relations
Rutgers University, School of Management and Labor Relations
University of Illinois at Urbana-Champaign, School of Labor and Employment Relations
Florida International University, College of Business
Pennsylvania State University, School of Labor and Employment Relations
Massachusetts Institute of Technology, Sloan School of Management/IWER

Organizational Members

American Federation of Teachers, AFL-CIO
Berkeley Research Group, Labor and Employment Practice
Boston University, Questrom School of Business
Brandeis University, Heller School for Social Policy and Management
Cornell University, Institute on Conflict Resolution
Cornell University, School of Industrial and Labor Relations
Federal Mediation and Conciliation Service
Florida International University, College of Business
Ford Motor Company
Kaiser Permanente
Massachusetts Institute of Technology, Sloan School of Management
Pennsylvania State University, School of Labor and Employment Relations
Portland General Electric
Rollins College, Department of Business and Social Entrepreneurship
Rutgers School of Management and Labor Relations
United Auto Workers International
United Steelworkers of America
University of California, Berkeley, Center for Labor Research and Education
University of California, Berkeley, Institute for Research on Labor and Employment
University of California, Los Angeles, Anderson School of Management
University of California, Los Angeles, Institute of Research on Labor and Employment
University of Illinois at Urbana-Champaign, School of Labor and Employment Relations
University of Minnesota, Center for Human Resources and Labor Studies
University of Toronto, Centre for Industrial Relations and Human Resources
U.S. Department of Agriculture
West Virginia University, College of Business and Economics
W. E. Upjohn Institute for Employment Research

For more information about Organizational Membership with the LERA, please visit the LERA website: http://www.lera.illinois.edu/about/OrgMemInfo.html.

ALPHABETICAL LIST OF AUTHORS

You are invited to become a member of

THE LABOR AND EMPLOYMENT RELATIONS ASSOCIATION

The Labor and Employment Relations Association (LERA) was founded in 1947 by a group who felt that the growing field of industrial relations required an association in which professionally minded people from different organizations could meet. It was intended to enable all who were professionally interested in industrial relations to become better acquainted and to keep up to date with the practices and ideas at work in the field. To our knowledge there is no other organization that affords the multiparty exchange of ideas we have experienced over the years–a unique and valuable forum. After 70 years, both our academic and practitioners agree with the conviction of the founders that the encouragement, reporting, and critical discussion of research are essential if our professional field is to advance.

Our local and national membership of 5,000 includes representatives of management, unions, government; practitioners in consulting, arbitration, mediation, and law; and scholars and teachers representing many disciplines in colleges and universities in the United States and Canada, as well as abroad. Libraries and institutions interested in the publications of the Association are also invited to become subscribing members and thereby receive the same member publications. Organizational memberships in the Association are also available.

Membership dues cover publications for 12 months and entitle members to the electronic newsletter, membership directory and proceedings of the annual meeting online, and printed copies of the annual research volume, and the magazine, *Perspectives on Work* (volumes also available online). Additional online resources include Chapter Profiles and an IR/HR Degree Program Listing for the United States, Canada, and Australia.

Dues for the 2020 calendar year are listed below. International memberships require an additional $25.00 per year to cover postage costs.

Regular Member...$199.00
Contributing Member $310.00
Emeritus Member (10 years+ membership and no longer employed) $110.00
Student Member (full-time, limited to 4 consecutive years) $30.00
Library Subscriber...$300.00

Tax-deductible financial contributions to the Association to support its educational activities are always welcome. You can make an online contribution at our website, leraweb.org, and choosing "Donate". If you are not already a member, we invite you to join the LERA by sending your membership application and dues payment. You can also join online at the LERA website, leraweb.org by choosing "Membership", or request an invoice. Inquiries regarding membership, meetings, and publications should be addressed to the LERA office.

LABOR AND EMPLOYMENT RELATIONS ASSOCIATION
University of Illinois at Urbana-Champaign
121 LER Building, 504 E. Armory Avenue
Champaign, IL 61820 USA
Phone: 217-333-0072 Fax: 217-265-5130

Email: LERAoffice@illinois.edu Web: leraweb.org

Also in the
Labor and Employment Relations Association Series

Employment and Disability: Issues, Innovations, and Opportunities
Edited by Susanne Bruyere $34.95

No One Size Fits All: Worker Organization, Policy and Movement in a New Economic Age
Edited by Janice Fine, Linda Burnham, Kate Griffith, Minsun Ji, Victor Narro,
Steven Pitts, and Teyo Reyes..$29.95

The Contradictions of Pension Fund Capitalism
Edited by Kevin Skerrett, Chris Roberts, Johanna Weststar, and Simon Archer $29.95

The Evolving Health Care Landscape: How Employees, Organizations, and Institutions are Adapting and Innovating
Edited by Ariel C. Avgar and Timothy J. Vogus.. $29.95

Inequality, Uncertainty, and Opportunity: The Varied and Growing Role of Finance in Labor Relations
Edited by Christian E. Weller..$29.95

The Disunited States of America: Employment Relations Systems in Conflict
Edited by David Jacobs and Peggy Kahn...$29.95

Collective Bargaining Under Duress: Case Studies of Major U.S. Industries
Edited by Howard Stranger, Ann C. Frost, and Paul F. Clark..$29.95

Public Jobs and Political Agendas: The Public Sector in an Era of Economic Stress
Edited by Daniel J. B. Mitchell..$29.95

Employee Ownership and Shared Capitalism: New Direction in Research
Edited by Edward J. Carberry..$29.95

Transforming the U.S. Workforce Development System: Lessons from Research and Practice
Edited by David Finegold, Mary Gatta, Hal Salzman, and Susan J. Schurman...............................$29.95

Book orders and request for prices and order forms for the above LERA publications should be sent to:

ILR Press, Cornell Univ., P.O. Box 6525, Ithaca, NY 14851
Phone 607-277-2211, Fax 607-277-6292
Web www.cornellpress.cornell.edu
Additional volumes are available through ILR Press.

Information about ordering printed copies of the *Proceedings of the LERA Annual Meeting* can be found online at https://lera.memberclicks.net/publications. Orders for printed copies of *Perspectives on Work* and other LERA publications should be sent to the LERA Office:

LERA, 121 LER Building, 504 E. Armory Ave., MC-504, Champaign, IL 61820
Phone 217-333-0072, Fax 217-265-5130
Email LERAoffice@illinois.edu, Web leraweb.org

Labor and Employment Relations Association (LERA)
University of Illinois at Urbana-Champaign
121 Labor and Employment Relations Bldg.,
504 E. Armory Ave.
Champaign IL 61820
USA

ISBN 978-1-7138-0276-1

9 781713 802761

CURRAN ASSOCIATES INC.
proceedings
.com

Land Use and Climate Change Impacts on Erosion and Sediment Transport

Climate Change Impacts on Sediment Dynamics: Measurement, Modelling and Management

ICCE Symposium 2018

Proceedings of the International Association of Hydrological Sciences (PIAHS) Volume 381, 2019

Moscow, Russia
27-31 August 2018

IAHS AISH